FLORIDA HONEY

and

ITS HUNDRED USES

By

WALDO HORTON, M.D.
and
ISABELLE S. THURSBY

Combined With

BEEKEEPING

By

J. J. WILDER

FLORIDA
Department of Agriculture

Creative Cookbooks
Monterey, California

Florida Honey and its Hundred Uses
by Waldo Horton and Isabelle S. Thursby

Beekeeping
by J. J. Wilder

for Florida Department of Agriculture

ISBN: 1-4101-0768-X

Copyright © 2004 by Fredonia Books

Reprinted from the original edition

Creative Cookbooks
An Imprint of Fredonia Books
Monterey, California
http://www.creativecookbooks.com

All rights reserved, including the right to reproduce this book, or portions thereof, in any form.

HONEY AND ITS HUNDRED USES

Contents

		Page
1.	Honey Production	4
2.	Honey Facts	8
3.	Minerals in Dark Honey	9
4.	Florida Honey	12
5.	Kinds of Honey	13
6.	Honey Cookery	20
7.	Honey Recipes	21
8.	Honey Bees and Their Products	63
9.	The Mystery of Sweets	65
10.	Honey and Nutrition	68
11.	Florida Laws Relating to Honey	70
12.	Tupelo Honey	72

BEEKEEPING

1.	Foreword	81
2.	Beekeeping in Florida	82
3.	Florida Laws Relating to Beekeeping	97
4.	Pollination of Subtropical Fruit Plants	99
5.	Honeybees in Florida's Pasture Development	103
6.	Honey Processing	107

Honey Production

Honey and Beeswax - Florida and U. S.

1956—REVIEW AND HISTORICAL DATA

The production of honey in Florida in 1956 amounted to nearly 17.4 million pounds valued at $3.3 million. This production pushes Florida to the third ranking State in the United States this year— behind California's 29.0 million pounds and Minnesota's 19.3 million. In value of honey and beeswax, California ranks first with $4,553,000, Florida second $3,476,000 and Minnesota third, $3,346,000. The State's 248,000 colonies yielding 70 pounds per colony, produced 17,360,000 pounds of honey and 330,000 pounds of beeswax.

Prices for Florida honey were in general a little higher in 1956 over 1955, bringing the value of honey to $3,298,000 and beeswax to $178,000. The average price received for honey for all methods of sale in 1956 was 19.0 cents per pound compared with 18.3 cents in 1955. Beeswax was 54 cents per pound, up 3 cents from 1955.

Year	Colonies of Bees	Pounds Honey Production per Colony	Pounds Honey Production	Value of Honey Produced	Value of Beeswax Produced	Total Value Honey & Beeswax
1945	182,000	50	9,100,000	1,447,000	71,000	1,516,000
1946	191,000	80	15,280,000	2,903,000	128,000	3,031,000
1947	195,000	42	8,190,000	2,097,000	72,000	2,169,000
1948	199,000	41	8,159,000	1,624,000	60,000	1,684,000
1949	189,000	60	11,340,000	1,678,000	68,000	1,746,000
1950	208,000	78	16,224,000	2,304,000	116,000	2,420,000
1951	218,000	82	17,876,000	2,842,000	140,000	2,982,000
1952	227,000	75	17,025,000	2,707,000	120,000	2,827,000
1953	238,000	76	18,088,000	3,057,000	124,000	3,181,000
1954	238,000	74	17,612,000	3,135,000	127,000	3,262,000
1955	238,000	55	13,090,000	2,395,000	114,000	2,509,000
1956	248,000	70	17,360,000	3,298,000	178,000	3,476,033

Honey production in the UNITED STATES in 1956 totaled 215 million pounds—19 per cent below 1955 and the smallest crop since 1948. The crop was produced by 5,332,000 colonies of bees, about the same number as in 1955; but the production per colony of 40.2 pounds was below the 1955 production at 47.8 pounds per colony. In mid-December producers had about 49 million pounds of honey on hand for sale—about 23 per cent of the estimated 1956 production. Beeswax production totaled 4,118,000 pounds, compared with 4,599,000 pounds in 1955.

Production of honey was below 1955 in all regions of the country except the South Atlantic and West where it was up 36 percent and 1 percent respectively. Spring and summer weather in the East North Central States was cool and wet and prevented bees from working much of the time resulting in the smallest crop since record began in 1939. Colonies from package bees built up slowly in this area. Drought conditions over much of the West North Central area resulted in a crop well below that of the previous year.

Honey production per colony averaged 40.2 pounds, compared with 47.8 pounds in 1955 and the average of 43.6 pounds.

Estimated stocks of honey on hand for sale by producers in mid-December totaled 49 million pounds—25 percent of production.

Beekeepers received an average price of 19.0 cents per pound for *all honey* sold in 1956, including the combined wholesale and retail sales of extracted, chunk and comb honey. This was about 1.2 cents higher than 1955, and the highest since 1947. These include sales by large and small apiaries owned by farmers and non-farmers. Price increases over 1955 were small but were recorded for all types of sale in most areas. Extracted honey in wholesale lots, the principal method of sale, brought an average of 15.2 cents per pound compared with 14.3 in 1955. Retail prices for extracted honey averaged 24.8 cents, 1.4 cents more than a year earlier. Prices received for chunk honey averaged 27.0 cents for sales in wholesale lots and 33.4 cents for sales in retail lots, each 0.4 cents higher than in 1955. Prices received for *Comb* honey sold in wholesale lots in 1956 averaged 31.8 cents, compared with 30.9 cents a year earlier, and retail prices average 36.4 cents, compared with 35.5 cents in 1955. Prices received by beekeepers for beeswax increased sharply in all areas during the year and averaged 54.6 cents per pound, compared with 51.2 cents in 1955.

HONEY PRODUCTION—IN POUNDS
According to rank in 1956 (000-Omitted)

State	1949	1950	1951	1952	1953	1954	1955	1956
California	21,900	22,550	28,246	48,974	23,628	33,831	30,072	29,044
Minnesota	24,708	23,375	23,375	24,795	21,335	19,440	24,300	19,280
FLORIDA	11,340	16,224	17,876	17,025	18,088	17,612	13,090	17,360
Iowa	17,974	17,302	10,890	17,072	14,091	10,360	18,975	12,400
Idaho	9,018	7,434	9,568	6,764	6,586	9,048	7,080	8,850
Wisconsin	12,675	12,025	14,550	15,908	15,600	10,812	18,512	5,036
Michigan	9,699	9,984	10,120	8,575	8,100	6,660	9,250	6,336
Texas	13,373	15,850	9,424	10,944	9,636	7,560	11,782	6,233
Georgia	4,104	4,305	5,160	5,580	4,598	4,100	2,412	6,210
Montana	3,596	3,720	5,200	5,146	6,030	5,780	6,188	5,934
OTHER STATES	98,591	100,184	123,707	111,848	96,722	91,881	112,125	94,841
TOTAL U. S.	226,978	233,013	258,116	272,641	224,414	217,084	253,786	214,524

FLORIDA BEEKEEPER PROGRESSIVE

To improve the marketing of the Florida honey crops there has been organized a cooperative marketing association with headquarters at Umatilla. At Umatilla this organization has a modern assembling, grading, processing, standardizing and packing plant. One of the ideas of the organization is to bring together the various farmer packs of barrel honey where it can be mingled and blended into a uniform product, and then it can be marketed more effectively. The plant has been recently expanded and improved; this is a sign of its growth and popularity. Honey has been packed for a large exporter in New York City.

In reporting on the land of Canaan, the Bible says that when the children of Israel sent some scouts across the river to see what kind of place they were to move into, the scouts reported on their return that it was a land of "milk and honey." In their estimation no greater praise than that could be given a country. This land must have been a cattle growing and fruit raising country, and in recent years Florida has become a cattle and fruit country.

Florida has a great variety of honey nectar producing plants and a long growing season. Perhaps no name among the honey varieties has more appeal than "Orange Blossom Honey."

The Tupelo honey, because it will not crystallize, has a wider and more varied use than most other honey. There are many varieties of honey, and all are important, that cannot be given in a short story, however some will be named just to show the diversity. For example, Palmetto and Galberry produced almost over the entire state, Ti Ti in West Florida and Mangrove and Sunflower in the southern part of state. Surely Florida is fast becoming the "land of milk and honey," for as citrus plantings expand and forest fires are controlled more honey can be produced.

Honey Facts

Honey is wholesome, natural food.

It keeps indefinitely, if stored in a warm, dry place.

It gives sweetness plus flavor.

It may wisely be substituted for sugar or molasses.

It is a highly energy giving food; especially easily assimilated.

It contains small amounts of mineral matter and vitamins.

It possesses slight laxative properties and helps many with constipation.

For those OVER-WEIGHT, used moderately, it gives some sweet without fear of the heavy fat production of cane sugar.

It is an ideal milk modifier (plus water) for infants.

It will not harbor bacteria and will actually kill them (by hygroscopic action).

Most all pure honeys granulate in time, some hard, some 'mush like'; beat or 'work' a granulated honey and you make a delicious fine-grain 'spread' of it. Any granulated honey can be reduced to its original consistency and flavor by heating in a waterbath at 125° F. for a half hour or more. Heating above 130°F. removes some of its delicate aroma and flavor.

Minerals in Dark Honey

From the Chemistry Division of the University of Wisconsin have come numerous papers which are of much interest to beekeepers. During the past twenty years a number of these papers dealing with the oils found in horsemint have received attention in the Beekeepers Item. In 1932 Kathora Remy of San Antonio read a short article on the chemistry of honey before the Texas Academy of Science. This paper likewise was mentioned in the Item. In 1932 Dr. H. A. Schuett and Kathora Remy published a paper in the Journal of the American Chemical Society* that had to do with the color of honey and the mineral content. They showed that from analysis of many samples of honey originating in numerous honey producing sections in the United States and Hawaii it was very probable that the darker the color of the floral honey the higher the percent of mineral content. In their analyses they show the varying amounts of silica, iron, copper, and manganese. Based on the statements made in the paper of Schuette and Remy the Item editorially mentioned the fact that in all probability it was the mineral content in the darker honey which had made these honeys more popular with people that use large amounts of honey.

In 1937 Schuette and Huenink in "Food Research"* published an extensive article in which the percents of silica, phosphorus, calcium, and manganese found in light and dark honeys are contrasted and commented upon. In this paper notes as to the varying amounts of these minerals in various beekeeping localities throughout the world are discussed. In the summary of this the authors say:

"In an examination of 35 samples deemed to include representatives of most of the honeys produced commercially in the United States it has been found that there apparently exists a qualitative relationship between degree of pigmentation, as revealed by the present-day practice of color-grading this food, and mineral content."

In January 1938 in the same Journal* Schuett and Triller presented a paper telling of sulphur and chlorine as it is found in honey. Numerous samples of honey from representative honey producing districts were analyzed and the same apparent relationship between the minerals sulphur and chlorine and the degree of color was rather definite. There was no general uniformity of occurrence of sulphur and chlorine through the samples utilized.

Investigations in beekeeping have largely taken the mystery out of the beehive. The chemists are at the present time taking the mystery out of honey and are establishing a very definite reason why honey is one of the best of foods. The earliest investtigations were relative to the enzymes, invertase, diastase, and catalase. As later chemical analyses of foods indicated that small percent of certain elements were necessary for a proper balance in the foods, the food chemists have ceased to investigate the very complex organic chemical compounds and turned their attention to the inorganic material found in honey. To the ordinary beekeeper the method of finding out how much sulphur or iron is contained in honey is a greater mystery than the hidden activity within the beehive is to the reading public. For the benefit of the beekeeper whom we know is much interested in knowing how these discoveries are made, this explanation is given.

The samples of honey are secured. Then the honey is freed from water and all organic compounds by means of heat. The remainder called ash is as a rule a white fluffy substance. This substance which is very small in proportion to the honey used is weighed on the most delicate of scales and then placed in solution in various liquids. By the addition of solutions of other chemicals and watching the changes which take place the chemist is able to tell to a very accurate degree the exact mineral and the amount thereof which is found in that portion. Then by combining the findings of all portions the total mineral content of the honey is obtained. Beekeepers appreciate the work of the chemists as they know that work of this kind gives definite information that can be depended upon in the formulation of sales talks.—H. B. PARKS, San Antonio, Texas.

---

*Schuette, H. A., and Triller, Ralph E., 1938. Mineral constituents of honey. III. Sulfur and Chlorine. Food Research 3, 543.

SCIENTISTS FINDING NEW USES FOR HONEY

The power of honey to absorb and retain moisture gives it many industrial uses, in addition to its value as food, studies by the Bureau of Chemistry and Soils show. This quality of honey, called "hygroscopicity," will make for greater use of the honey grades not adapted to home use.

Bureau studies included the behavior of honeys of different flower origin—white clover, tupelo, buckwheat, tulip poplar, and mesquite. All these honeys are found useful in commercial baking of bread, cake and cookies. When these products are made with part honey in place of sugar, they lose less moisture after being stored 7 days than bread, cake, and cookies made with other sweetening agents. Buckwheat honey gives particularly good results.

Honey is also useful in candy making. It is suggested for curing tobacco, in the same way that sugar and maple sugar are used. Among other industries that offer outlets for comparatively large quantities of honey are brewing, wine making, and vinegar manufacture.

Florida Honey

By Dr. WALDO HORTON

Nature's Own Sweet—Nature's Oldest Sweet

Chemistry is now corroborating experience and proving that our honey from sub-tropical and tropical plants contains more minerals and is more health-giving. We have more variety than almost any state, to please those who like a change of flaxor. For those who like it standard and always the same a Florida blend is recommended.

FOOD VALUE

In infancy milk is a balanced and sufficient food. As we become more active a higher calorie addition becomes necessary. But foolish and taste-tickled mankind has gone too far with varieties and mixtures. Sane thinkers are now reverting to the more simple. In this very generation we are sure to see increasing thousands going back to the more elemental, natural foods. The Biblical recommendation of milk and honey, (Num. 13-27; Gen. 43-11; Ps. 19-10; Math. 3-4; Is. 7-15), should again be taken seriously.

Honey is a monosaccharide sugar, chiefly fruit sugar. This sugar is the natural end-product of digestion, so that honey is already digested and easily assimilable. It is sweeter than cane sugar but also contains more water and the amount varies in honeys from different flowers.

There are many kinds of honey, almost as many as there are different flowers, though some flowers do not produce nectar (honey). Bees gather the nectar, and in the hive process it and store and condense it in the comb as honey. Extracted honey is thrown out of the comb by an extracting machine and strained, and is used on the table and in cookery like syrup.

Honey adulterated with cheap syrup is not so common as thought, on account of the rigid Pure Food Law, but if there is real reason to suspect this adulteration, a sample sent to Gainesville or Washington will disclose the truth.

CARE

Honey absorbs atmospheric moisture, granulates rapidly if cold; hence keep it in a warm dry place where you would keep salt.

Keep under tight cover; insects like it, too.

Do not keep in refrigerator! (Perhaps comb honey, a short time.)

Granulated honey is not spoiled honey; in fact nature does that to preserve it. Some people like granulated honey. If you wish it liquid like new, heat in waterbath at 125° or 130°F. for an hour.

Comb honey is hard to keep prime here for many weeks outside of beehive. (65°F. dry storage is needed.)

Remember good honey properly kept does not spoil and is still delicious when a year or two old. (A few careless drops of water or impurity may make it spoil.)

Before serving thick extracted honey, set container in warm water a few minutes; this makes it pour more easily.

Honey, being imperishable, can be purchased in large quantities and stored.

KINDS OF HONEY

Saw Palmetto Honey

This is Florida's most universally produced and used honey. It usually grades amber color, sometimes dark amber and occasionally light amber; all becoming darker with age. Its mild flavor and odor are characteristic and pleasing. Medium body.

Because its source-plant is used somewhat in medicine, it is thought by many to be unusually health-giving. It granulates slowly. Use for both table and cooking. Much Florida honey found in our stores has at least some of this mixed in by the bees.

SAW PALMETTO (Serenoa)
What native Floridian, but thrills at this? Just hear the hum of the bees!

Cabbage Palm Honey

A thin bodied, light amber honey of very mild flavor and odor. Excellent for cookery and sweetening drinks where mild flavor is desired.

Tupelo Honey

This is produced from the tupelo gum tree (Nyssa) which grows along the streams of West Florida. It is light amber in color, of heavy body and mild flavor. It has the most varied use of Florida honeys, having been tried scores of ways and not found wanting. It does not granulate; hence is much sought for by packers to blend with other honeys to keep down their granulation.

Orange Blossom Honey

Makes us think of weddings and the perfume-laden air of springtime. In all the kingdom of beedom what sweeter words than Orange Blossom! To stand in an orange grove and watch these little workers hustle from blossom to blossom makes one realize that they too, regard it the choicest of nature's golden sweets.

In cooking and candy-making few honeys carry over so much distinct flavor. At the fountain, in the tea room, as well as the diet kitchen, its exquisite possibilities have yet been scarcely thought of.

Because of its peculiar distinction it is much counterfeited. As many as twenty different mixtures, colors, and flavors have been called orange blossom honey. Genuine orange blossom honey is light amber in color, heavy in body, has the real aroma of the grove in bloom and does not darken or change flavor much with age. In aging it granulates readily.

Gallberry Honey

This honey is produced from the gallberry bush (Ilex glabra), which grows in flatwoods sections and blossoms usually in May. It is almost a water white honey, with a heavy body and very mild flavor and is considered one of our finest honeys. Due to the damage done this plant by burning the woods, very little gallberry honey has been produced in recent years in Florida. It is almost too fancy a honey to use in baking, but is wonderfully adapted for icings, ice cream and for direct sweeting in other desserts where mild flavor is desired.

CABBAGE PALM (Sabal)
Walk near these blooms in July and you may think you have discovered a swarm of bees. It is only normal industry working the many thousands of tiny blossoms.

Mangrove Honey

From the salt marshes of South Florida come large quantities of another of our 'best' honeys. Black mangrove (Avic. nitida) produces a delicious flavored honey almost as light colored as gallberry, light in body but unusually sweet, due to a large content of dextrose.

Holds an enviable place with many devotees and gaining popularity fast.

Other Commercial Florida Honeys

These are Wild Sunflower from the Everglades region, a delicious fall honey of amber or light amber color and good body; Partridge Pea, which is a darker, stronger product, excellent for cooking and baking; and Goldenrod, a popular fall honey.

Besides these nine, Florida produces over a score of others, but rarely distinct or in pure state enough to be seen commercially.

Sub-tropical honeys are rich in minerals and vitamins!

Much Florida honey comes from the flowers of wild trees, shrubs and small plants. Among these are the tupelo tree, two varieties of palmetto, mangrove, magnolia, ti ti, gallberry, gopher apple, chinkapin, and a few less important.

Among the cultivated trees and plants that yield nectar to the honey bee are: citrus trees, clovers, pennyroyal, partridge pea, watermelon, etc.

The ti ti is a shrub or tree of swamps of North Florida with an exquisite bloom much adored by the bees. The honey is light and mild.

SOME HONEY PLANTS OF FLORIDA

Common Name	Botanical Name	Months of Year in Bloom	Localities Where Found
1. Saw Palmetto	Serenoa serrulata (Michx.) Hook	May and June	Practically all over the State
2. Black Mangrove	Avicennia nitida, Jacq.	June and July	Around ocean's edge from New Smyrna to Tampa Bay
3. White Tupelo Gum	Nyssa aquatica L.	April and May	Along rivers and overflow land in western part of State
4. Partridge Pea	Chamaecrista spp.	June, July, August and September	Throughout sand ridge section
5. Gallberry	Ilex glabra (L.) A. Gray	April and May	Throughout flatwoods section
6. Wild Sun Flower	Helianthus spp.	November and December	Southern part of State, principally around Lake Okeechobee
7. The Summer Fair Well	Kuhnistera pinnata (Walt.) Kuntze	September and October	On light, sandy, well drained soil throughout the State
8. The Wonder Honey Plant	Pentstemon Pentstemon (L.) Britton	April, May, June, and July	Along the coast around Apalachicola Bay
9. Black Tupelo Gum	Nyssa biflora Walt	March and April	Along streams in the western part of the State

10. Spring Ti Ti	Cyrilla parvifolia Raf.	February and March	In Western part of State along small streams and bay heads
11 Pennyroyal	Pycnothymus rigidus (Bart.) Small	December, January and February	Southern part of the State
12 Cabbage Palmetto	Sabal Palmetto (Walt.) R. & S.	July	Along the coast, through the hammocks and along the lakes
13 The Pepper Bush	Clethra alnifolia	July and August	Throughout flatwoods section
14. Mexican Clover	Richardia scabra St. Hil.	July, August and September	In many cultivated fields throughout the State
15 Goldenrod	Solidago spp	October and November	Throughout the State
16. The Snow Vine	Willugbaeya scandens (L.) Kuntze	July	Western part of the State
17 Gopher Apple	Chrysobalanus oblongifolia Michx.	May	Throughout sand ridge section
18 Blackberry	Rubus spp.	April and May	All over the State
19 Chinkapin	Castanea spp.	April and May	North and West Florida
20 Citrus	Citrus spp.	March and April	Throughout Central and South Florida with Satsumas in North and West Florida

HONEY COOKERY

By ISABELLE S. THURSBY

Honey is one of the oldest known human foods and was considered one of the choicest by the ancients. In those days honey was the nectar of the gods. And even today no food is more interesting than honey. The very name of honey carries an appeal possessed by no other food. There are many reasons why this delicious, natural unrefined, unmanipulated sweet should be used abundantly in the diet, not only in its natural state but as an ingredient of cooked food.

Many people think of honey primarily as a delicious spread for bread—hot biscuits, waffles and griddle cakes. But when included in cookery processes not only does it supply the sweetening, but its distinctive, individual flavor combined with the other ingredients, produces a delectable blend of flavor that not only is different but is intriguing as well.

The use of more honey in cookery is to be encouraged because of its superior flavor, food and health value and availability.

A new set of recipes is not necessary in order to use honey for one can substitute by following a few basic principles.

First: Remember that one cup of honey contains ¼ cup of liquid.

Second: Deduct ¼ cup liquid from the recipe when using 1 cup honey.

Third: Florida honey is very sweet, so no alteration need be made in the recipe regarding sweetening power, as one cup of honey is equal to one cup of sweeting. Liquid or granulated honey is equally satisfactory to use.

Fourth: Honey retains moisture to a greater extent in the product than does sugar. In making frostings this fact should be taken into account and the product should be cooked to a higher density than is done when using sugar.

Fifth: In using honey as the sweetening agent in the place of granulated sugar, the difference in composition and flavor must be considered.

HONEY RECIPES

Different honeys have definitely characteristic flavors and aromas, hence the flavor of any product made by a given recipe will vary with the kind of honey used. The milder honeys should be used for salads, fruit sauces, meringues and beverages, whereas the stronger honeys are perfect for gingerbread, spice cake, and for combinations that contain chocolate.

HONEY IN BAKING

Bread and honey for thousands of years have been recognized as a most acceptable food. Breads, cakes, cookies and waffles backed with a small amount of honey have a distinctive flavor that is very pleasing to most palates, and for those cakes and cookies where moist keeping is desired, honey is desirable.

Honey has long been associated with crisp, tender golden waffles. Now honey is often baked in them or, better and more delicious still, honey may be served as a sauce or paste by creaming together one part butter with 2 parts honey—beating smooth. Appetizing and satisfying are hot honey muffins, crisp and brown, spread with honey butter or honey in the comb. Honey pecan muffins are delicious for Sunday supper with chicken salad and honey. Whole wheat or oatmeal muffins are very popular with children. Cakes and cookies made with honey, baked when convenient, ready when needed, may be kept on hand constantly for use on busy days, or for surprise guests.

HONEY CREAM WAFFLES

1 egg beaten very lightly
4 tsp. baking powder
2 cups flour
1 tsp salt

2 cups milk
¼ cup butter or butter substitute melted
3 tbsp. honey

Mix shortening, honey and salt with beaten egg. Sift baking powder and flour together. Stir in alternately with flour and milk until full amount has been added. By using this regulation honey batter and adding nuts, candied or dried fruits, one may obtain a delicious result. Try also a honey pecan or a honey date waffle.

GALLBERRY (Inkberry) (Ilex galbra)
The berry itself may taste like gall and look like ink, but the bee takes wondrous Nature while at her best and gathers for her human friends from the chasteness of the bloom, one of the four finest honeys of Florida.

HONEY OATMEAL MUFFINS

1 cup milk
1½ cups flour
1½ cups oatmeal
1 egg

3 tbsp fat
1 cup honey
½ tsp. salt
3 tsp. baking powder

Mix dry ingredients, add milk, beaten egg, honey and melted fat, (slightly cooled). Mix but do not heat. Place in greased muffin irons. Bake in hot oven (400°F.) 30 minutes.

HONEY AND NUT BRAN MUFFINS

½ cup honey
1 cup flour
¼ to ½ tsp. soda
¼ tsp. salt

1 cup bran
1 tbsp. melted butter
1½ cups milk
¾ cup finely chopped pecans

Sift together the flour, soda and salt, and mix them with the bran. Add other ingredients, and bake for 25 minutes in a hot oven in gem tins.

HONEY BISCUITS

⅓ cup fat
2 cups flour
⅔ cup milk

½ tsp. salt
4 tsp. baking powder

Sift flour, baking powder and salt, add milk gradually and combine to a dough consistency. Pat out into a sheet ½ inch thick. Cream ¼ cup butter with ¼ cup strained honey. Use part of this mixture for spreading on the dough. Roll up and cut off like cinnamon rolls. Use the balance of the butter and honey mixture and spread thickly over bottom of pan. Arrange rolls, allowing ½ inch space around each. Bake in a hot oven (375° F.) 12 to 15 minutes. Cinnamon may be added to the butter and honey mixture and raisins or candied fruit may be chopped and sprinkled over the biscuit dough before rolling, if desired, or nut meats may be used in the same way.

MANGROVE TREE

HONEY OATMEAL BREAD

2 cups rolled oats
2 cups scalded milk or boiling water
1 yeast cake
4/5 cup flour
1 tsp. salt

½ cup honey
½ cup lukewarm water
2 tbsp. shortening
(optional—1 cup chopped pecans or candied orange peel)

Pour scalded liquid over the oats and shortening. Cover and let stand until lukewarm. Dissolve yeast cake in the warm water, add honey and stir into the oatmeal. Add 1½ cups flour, beat well, cover and allow to rise for 1 hour until light. Then add the salt, the rest of the flour and the nuts or candied peel and enough flour to make a dough and knead until smooth. Place in a greased bowl, cover and let stand again in a warm place until double in bulk. Shape into small loaves, put into well greased pans, filling them a little more than one-half full. Let rise to top of the pan and bake 50 minutes in a hot oven.

HONEY ORANGE GRAHAM BREAD

1 cup scalded milk
1 tsp. salt
⅓ cup lukewarm water
1½ cups graham flour
4 tbsp. honey

1 yeast cake
1½ cups bread flour
½ cup candied orange peel
½ cup pecan nut meats

Mix milk, honey and salt. When lukewarm add yeast cake dissolved in lukewarm water, and flour. Mix and then add orange peel and nuts, cut in small pieces. When thoroughly mixed, let rise until double in bulk. Shape into loaves in bread pan and let rise again until double its bulk. Bake in a 350° to 380° F. oven from 40 to 60 minutes. This mixture can be baked in muffin tins and served while hot.

HONEY NUT BREAD

¾ cup honey
1 egg
1 cup milk
3 cups flour

3 tsp. baking powder
½ tsp. salt
1 cup nut meats, chopped

Mix, put into a greased and floured loaf pan. Let stand about one hour. Bake in a slow oven for about 40 minutes or one hour.

GOLDENROD

HONEY SPONGE CAKE

1 cup cake flour
½ cup sugar
½ cup strained honey
5 egg whites
5 egg yolks

¼ tsp. salt
½ tsp. vanilla
¾ tsp. cream of tartar
2 tbsp, boiling water

Sift and measure flour and sugar. Beat egg yolks until thick and lemon colored. Add sugar and beat well; add honey and combine lightly. Add boiling water a tablespoon at a time. Beat ½ minute, add flavoring and flour and lastly fold in the beaten egg whites. Pour into a tube pan and bake for 50 minutes in a very moderate oven (300° F.). When baked, invert on cake cooler and allow to cool before removing from pan.

HONEY DOUGHNUTS

1 egg
1 cup honey
2 tsp. baking powder
Flour

1 cup sweet milk
2 tbsp. shortening
1 tsp. salt

Cream honey and shortening together, add the egg well beaten and the other ingredients. Mix well and add flour enough to roll out and cut easily. Fry in hot fat. The honey will keep these delicious doughnuts moist much longer than usual.

HONEY GINGERBREAD

½ cup fat
¼ cup sugar (brown)
½ cup sour milk
½ tsp. cinnamon
1 tsp. baking powder
1½ cups flour

¾ cup honey
1 egg
½ tsp. soda
⅛ tsp. cloves
¼ tsp. salt
½ tsp ginger

Sift dry ingredients. Cream fat and honey, add brown sugar, egg, sour milk and sifted dry ingredients. This will be a thin batter, but do not mind that. Bake in a well-greased pan for 25 minutes in a moderate oven (350° to 375° F.). This is a delicious gingerbread and may be kept for several days, reheating before serving. Serve with or without Honey Meringue icing.

HONEY MERINGUE (7 Minute Icing)

1 egg white ½ cup honey (strained or granulated)

Place honey and unbeaten egg white in top of double boiler. Cook seven minutes, beating with dover egg beater while cooking. Remove from double boiler, beat and spread as desired.

DATE BARS

1 cup honey
3 eggs
1 tsp. salt
1 tsp. vanilla

1 cup flour
1 tsp. baking powder
1 cup dates
1 cup nut meats

Beat the eggs well and add the honey, salt and vanilla. Mix and sift the flour and baking powder, add the dates and nuts (cut in small pieces), then combine with the egg mixture. Pour into a greased, shallow pan, spread one-fourth inch thick. Bake in a moderately hot oven 30 to 40 minutes. Cut in strips before removing from the pan. Store in a crock or cake box for several days, as the date bars improve after standing. Roll in powdered sugar before serving.

HONEY OATMEAL COOKIES

1 cup honey
⅔ cup fat
½ tsp. salt
2 eggs, beaten
2 cups rolled oats

2 cups flour
½ tsp. soda
2 tsp. baking powder
1 tsp. cinnamon
1 cup chopped raisins

Cream the fat and honey together, then add the eggs. Mix and sift the flour, soda, baking powder, cinnamon, and salt, and add to the wet mixture together with oatmeal. Dust the raisins with some of the flour and add them to the dough, mixing well. Drop by teaspoonfuls on a greased pan. Bake in a moderate oven 10 to 12 minutes.

BLACK MANGROVE (Avicennia Nitida)
Another botanical paradox of Florida is this shrub-like tree which grows with its feet in salt water (marshes) and produces large quantities of one of our most delicious sweets.

DROP COOKIES NO. 1

1 cup fat	½ tsp. salt
¼ cup sugar	½ tsp. vanilla
½ cup strained honey	2 drops almond extract
2 cups flour	½ cup nut meats
2 eggs	½ cup raisins
½ tsp. soda	

Cream fat and sugar thoroughly. Add honey, beaten eggs and flavoring. Sift flour, soda and salt together and add to first mixture. Combine with lightly floured nuts and raisins. Drop by dessert spoonfuls on oiled baking sheet. Bake in moderate oven (350° to 375° F.).

LEMON NUT DROP COOKIES NO. 2

½ cup butter	2 egg whites whipped
2 egg yolks, beaten	½ cup sugar
Grated rind one lemon	1 tsp. salt
3 tbsp. lemon juice	1 cup honey
3¼ cups pastry flour	Shredded coconut if desired

Cream the butter, beat in the sugar and add the egg yolks and lemon. Then stir in three cups of flour and the salt and soda sifted together, alternately with the honey. Fold in the beaten egg whites and stir in the nut meats, floured with the remaining fourth cup of flour. Drop by teaspoons onto a buttered baking pan two inches apart. Bake in a moderate oven (350° F.) from 15 to 25 minutes. Sprinkle with shredded coconut before baking, if desired.

HONEY NUT BROWNIES

¼ cup butter	½ cup sugar
2 ounces chocolate	½ cup flour, sifted with ¼ tsp. baking powder
½ cup honey	
2 eggs	1 cup chopped nut meats

Butter and chocolate should be melted together, then add honey, then flour and baking powder, then nuts. Bake 45 minutes in a slow oven. For immediate use it is better to use ½ sugar and ½ honey. Cut in strips one-half inch wide and 2 inches long. To pack away in a jar, use all honey instead of part sugar and do not use until after two weeks. Roll strips in powdered sugar before packing.

CORAL VINE (Antigonon)
A distant cousin of northern buckwheat—much liked by the bees. In larger plantings would help beautify our roadside fences, and produce another distinct honey. The same can be said of several other ornamental honey plants: Assonia, Yucca, Vitex.

CHOCOLATE REFRIGERATOR COOKIES

½ cup brown sugar
½ cup shortening
½ tsp. salt
1 tsp. baking powder
2½ cups flour
½ cup honey

1 egg
¼ tsp. soda
½ cup pecans
4 to 6 tbsp. cocoa, depending upon degree of chocolate flavor desired

Cream sugar, honey, shortening and egg. Add dry ingredients, then nuts, shape in a loaf or place in refrigerator cookie mold. Chill several days to allow sufficient ripening of dough. Slice off and bake in hot oven (400° F.) for about 12 minutes. After baking, if allowed to stand for several days, the cookies will improve in flavor.

HONEY FUDGE SQUARES

½ cup cocoa or 2 squares bitter chocolate
⅓ cup shortening
⅓ cup pecans, or black walnuts
¼ tsp. soda
Pinch salt
1 tsp baking powder
½ cup honey

½ cup brown sugar
1 cup chopped dates, or candied orange peel
1 egg
2 cups flour
¼ cup sour cream, or ¼ cup of evaporated milk to which has been added ¼ tsp. vinegar

Melt chocolate over hot water if squares of chocolate are used. Blend the melted chocolate or cocoa with honey, brown sugar and shortening. Add 1 egg, then sour cream. Add sifted ingredients. Then add the nuts and dates or peel. Spread batter to about ½ inch depth in flat pan and bake in moderate oven about 35 minutes. When cool, cut in squares.

HONEY COOKIES

1 cup honey
¼ cup butter
⅓ cup pecans cut in pieces
Grated rind of 1 lemon
½ tsp ground cinnamon

½ tsp ground cloves
½ tsp cardamon seed
2 tsp. baking powder
2¼ cups flour

Heat the honey and butter together for about 5 minutes; add all the other ingredients except the baking powder, and mix thoroughly. When somewhat cooled, sift in the baking powder and mix again. Let stand overnight. Roll thin and cut

LATE VALENCIA ORANGE TREES

into cakes of desired size and shape. Place on greased baking sheet or in shallow pan; if desired decorate with bits of citron and halves of almonds. Bake to an amber color (about 8 to 10 minutes at 350° F.).

HONEY COOKIES

½ cup butter
¾ cup sugar
1 egg and
1 egg yolk
½ cup honey

Grated rind of 1 lemon
3 cups flour
4 tsp. baking powder
1 egg white
Pecans, chopped

Cream the butter and sugar together and add the egg and egg yolk beaten together, the honey, lemon rind, and the flour sifted with the baking powder. More flour may be required. The dough should be stiff enough to be easily handled. Take a small portion of dough at a time, knead slightly, roll into a thin sheet and cut into cookies of any desired shape. Set the shapes on a greased pan. Beat the white of the egg (left for the purpose) a little; use it to brush over the top of the cookies in the pan, then at once sprinkle on some finely chopped pecans and a little granulated sugar. Bake in a moderate oven (about 10 minutes at 350° F.).

HONEY ORANGE CUP CAKES

⅓ cup butter
1 orange juice and rind
2½ cups pastry flour
2 tsp. baking powder
¾ cup honey

½ cup sugar
2 eggs well beaten
¼ tsp. salt
¾ cup broken walnut meats

Cream the butter and add the sugar gradually. Beat in the orange juice and rind and the eggs. Mix together the flour, salt and baking powder. Stir in the broken walnut meats and mix well. Add alternately to the cake mixture with the honey and bake in cup cakes 15 to 25 minutes at 350° F. If desired, ice with Honey Meringue Icing.

HONEY MERINGUE ICING (Boiled)

1 egg white
4 tbsp. water
Pinch of salt

1 cup honey
1 tsp cream of tartar

Combine all ingredients and cook slowly over low heat or in a double boiler, beating constantly until mixture stands up in

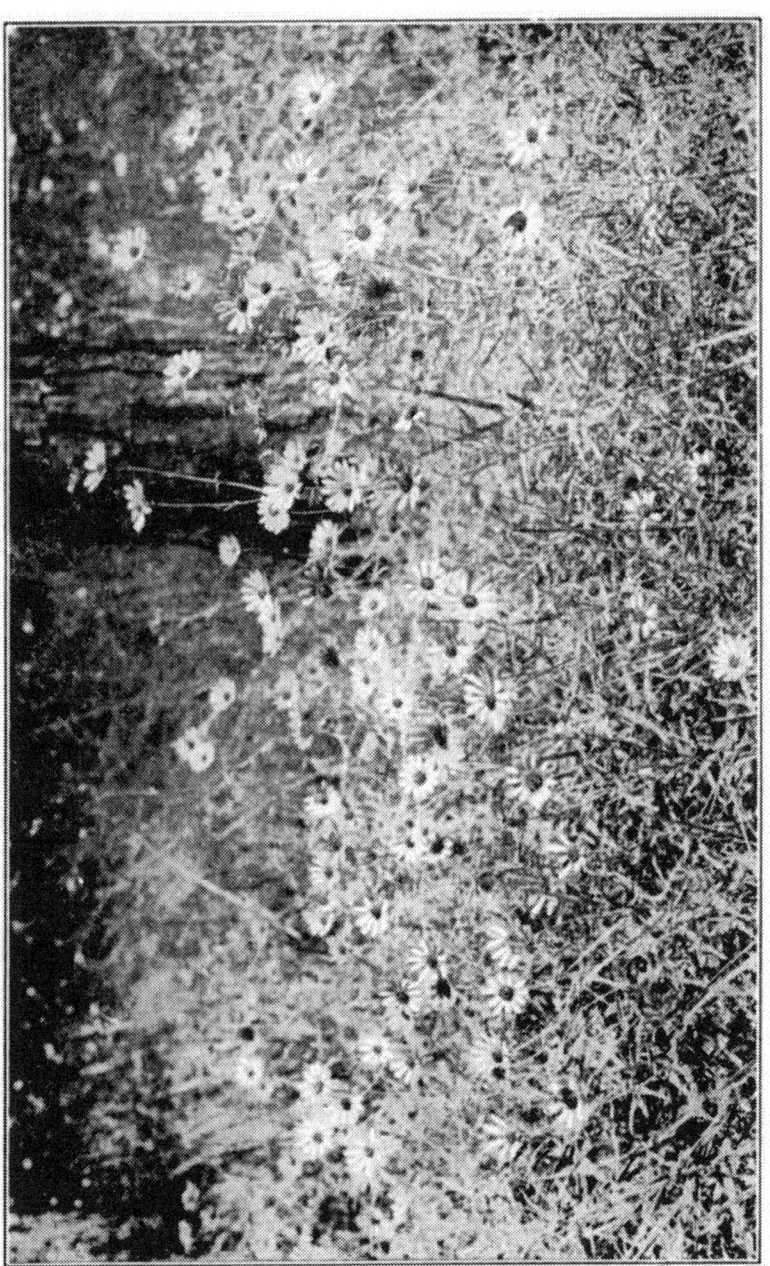

FLORIDA SUNFLOWER (Helianthus Floridus)
Grows in many parts of the State, but in Everglades Region, takes the secretion of nectar seriously and much fine honey is the result.

peaks. It may be beaten until creamy when removed from heat. This is a delicious meringue topping. It does not set on the outside, but is creamy and fluffy.

SPICED JELLY ROLL

3 eggs	1 cup flour
⅜ cup sugar	1 tsp. baking powder
¼ cup water minus 1 tbsp.	¼ tsp. salt
½ tsp. vanilla	¼ tsp. cloves
⅜ cup strained honey	1 tsp. cinnamon
	2 tbsp. melted butter

Beat yolks, add sugar, honey, water and vanilla. Sift flour, baking powder, salt, and spices, and add to first mixture. Add melted butter and fold in egg whites. Bake in shallow pan lined with well oiled paper in a hot oven (375° F.) for 20 minutes. When baked, invert on a cloth dusted with powdered sugar. Remove paper, trim off edges, spread with spiced roselle or blackberry jam. Roll cloth around cake and allow to "set" for a short time.

PECAN HONEY CAKES

4½ cups flour	2 eggs
1 tsp. salt	1 cup strong coffee
1 tsp. soda	¼ lb. sliced candied citron
1 tsp. cinnamon	¼ lb. sliced candied orange
1 tsp. cloves	or grapefruit peel
1 tsp. allspice	¼ lb. sliced guava paste
2 tbsp. cocoa	½ cup shortening
1 cup honey	1 cup brown sugar
	1½ cups chopped pecans

Sift flour, salt, soda, spices and cocoa together. Mix the chopped nuts and sliced fruit peel through the flour with the finger tips. Cream the shortening; stir in the sugar gradually. Add the beaten eggs and honey. Stir in the fruit and flour mixture alternately with the coffee. Spread the mixture on well-oiled baking sheets or shallow pans, making a layer about ½ inch thick. If baking sheets are used, leave a space about 1½ inches wide at the open end to allow for spreading. Bake in a moderate oven (350° F.) for 20 to 30 minutes.

The hot cake may be spread with thin layer of icing made by stirring lemon juice into confectioners sugar (3 to 4 tbsp. lemon juice for 2 cups sugar). Cut in 2-inch squares when cool.

Store in a tightly covered box for at least one week. Yields about 120 squares.

ORANGE HONEY COCONUT CAKE

½ cup shortening
½ cup sugar
½ cup orange honey
5 egg yolks
1¾ cups all-purpose flour
3 tsp. baking powder
½ tsp. salt
⅝ cup milk
1 tbsp. grated orange rind
1 tbsp. orange juice

Cream shortening; add sugar gradually and cream well. Add honey and mix well; add the very well beaten egg yolks. Sift flour once before measuring. Sift flour, baking powder and salt together. Add to creamed mixture alternately with the milk. Add orange rind and juice. Bake in well greased and floured pan for 40 minutes in moderate oven (350° F.). Ice with Honey Coconut Meringue.

HONEY COCONUT MERINGUE ICING

⅓ cup honey
1/16 tsp. salt
2 egg whites
½ cup toasted coconut

Heat honey to 240° F., or until it spins an 8-inch thread. Pour slowly into stiffly beaten egg whites and beat with egg beater constantly. Add salt and continue beating until mixture is fluffy and will hold shape.

Spread on warm cake and sprinkle top with the coconut, lightly toasted. Place pan of cake on board or in another pan to prevent further browning and return cake to oven to set meringue. Bake 10 minutes in very slow oven.

To toast coconut: Place 1 package coconut and 2 tsp. butter in pan and toast very slowly in oven, stirring frequently to prevent burning.

HONEY CITRON NUT CAKE

½ cup shortening
4 egg whites
¾ cup water or milk
4 tsp. baking powder
2¼ cups flour (sifted twice before measuring)

¾ cup honey (mildly flavored)
¾ cup sugar (white)
1 cup sliced citron
1 cup chopped pecans

Blend shortening, honey and sugar to a cream; add liquid and flour in which baking powder and salt have been sifted. Stir only until mixed and then add nuts and citron, folding in lastly the stiffly beaten egg whites. Pour into layer cake tins or flat oblong pan lined with waxed paper. Bake in moderate oven (350° F.) for 45 minutes to 1 hour, depending on depth of cake.

Other fruits or nuts may be used such as preserved watermelon rind or candied orange peel. Ice with Honey Icing.

HONEY ICING

2 cups powdered sugar
4 tbsp. heavy cream
Enough milk to give good spreading consistency

¼ cup honey
2 tbsp. melted butter
Citron slices to decorate cake

Blend butter and honey; add heavy cream and salt. Blend with powdered sugar and add just enough milk to give spreading consistency. Ice cake and decorate with citron slices. Put iced cake in cake box for two or three days before using. This cake may be kept from two to four weeks before using, as the honey keeps it moist and fresh and improves the flavor.

HONEY SPICE CAKE

1 cup shortening
⅓ cup sugar
1½ tsp. cinnamon
3 cups pastry flour
1 cup sour milk
1 tsp. vanilla flavoring

½ tsp. salt
¾ cup strained honey
½ tsp. cloves
4 tsp. baking powder
½ cup nut meats (broken)
½ tsp. soda
2 eggs

Cream shortening and add the sugar. Beat in the honey. Beat the yolks of eggs and add. Sift dry ingredients. Add ¼ cup to nuts and add these to the mixture. Add the remaining dry ingredients alternately with sour milk and vanilla. Fold in the beaten whites. Bake in a well greased loaf pan in a moderate oven (350° F.) for 45 minutes.

FLORIDA HONEY FRUIT CAKE

1 cup shortening	½ cup coffee
3 eggs	1½ cups pecans
½ cup citron	2 cups honey
½ cup candied gingered watermelon rind	3 cups flour
	¼ tsp. each cloves, salt, nutmeg and allspice
½ lb figs	½ tsp. soda
½ cup honeyed orange strips or honey orange marmalade	1 tsp. cream of tartar
	¾ tsp. cinnamon
¼ cup prunes	¼ cup candied pineapple
½ lb. dates	1 lb. raisins

Run figs, prunes, dates through food chopper. Add candied orange peel and raisins. Over this pour the honey and let stand from four days to a week.

Shred pineapple and citron. Sift dry ingredients, reserving ½ cup flour to mix with nuts, watermelon rind and pineapple. After the fruit and honey mixture has stood long enough, cream shortening and add to honey fruit mixture. Add the beaten eggs, then sifted dry ingredients, coffee and the floured nuts, pineapple, citron and gingered watermelon rind shreds.

Bake slowly (225° F.) for three hours if in one-pound tins. If the entire mixture is baked in one cake (five pounds) bake from four to five hours, depending on the depth of the cake. Brush top of cake with warm honey, wrap in heavy waxed paper, pack away in covered crock for at least a month. Before wrapping in cellophane for gift mailing or before serving, decorate top with honeyed orange peel, pecans, citron or pineapple. Yields five pounds fruit cake.

HONEY PIES

Pies have never lacked in popularity and made the honey way are of especially fine flavor and are good hot or cold.

HONEY PECAN PIE

2 tbsp butter	¼ tsp. salt
3 eggs	1 cup pecans, broken, depending on sweetness and richness desired
¾ to 1 cup honey	

Beat eggs slightly. Add honey and butter warmed and salt. Mix well, put in partly baked pie shell and bake in a moderate oven about 35 minutes.

HONEY PUMPKIN PIE

1½ cups steamed and strained pumpkin	1 cup honey
1 cup cream	1 tsp. cinnamon
1 cup milk	3 eggs, well beaten

Mix ingredients in order given and bake in one crust. Top with honey merinque. Or garnish each piece with a mound of whipped cream with honey in its center.

HONEY APPLE PIE

Make an apple pie as usual, but do not use any sugar after the apples—just the butter and cinnamon, and do not use a top crust. After it is baked, drizzle ½ to ⅓ cup honey over the apple filling and sprinkle one-half cup pecan pieces and let stand until apples become soft and absorb all the honey. Pears, peaches, loquats are all delicious used in the same way.

HONEY CREAM PIE

½ cup honey	1½ cups milk
4 tbsp. flour	2 egg yolks
¼ tsp. salt	1½ tbsp. butter

Blend flour with a part of the liquid (cold) until it is smooth. Add salt, honey and remainder of the liquid. Cook in a double boiler until thick, stirring frequently. Slowly pour a part of this cooked mixture over the beaten egg yolks, stirring constantly. Return to the double boiler and beat until the egg is cooked. Lastly add the butter. Pour this filling into a previously baked pastry shell. Cover with a meringue made from the two egg whites slightly sweetened with honey. Brown the meringue in the oven.

HONEY LEMON PIE

¾ cup honey
8 tbsp. flour
½ cup cold water

1 cup boiling water
1 lemon, juice and grated rind
2 egg yolks
½ to 1 tbsp. butter

Blend the flour and cold water until smooth; add the honey and grated lemon rind; slowly add the boiling water, stirring constantly. Cook in a double boiler until thick. Stir in the lemon juice. Slowly add part of this cooked mixture to the beaten yolks, stirring constantly. Return to the double boiler and heat until the egg is cooked. Lastly, add the butter.

Pour this filling into a previously baked pie crust and cover with a meringue made from the two egg whites slightly sweetened with honey and flavored with a drop or two of lemon extract. Brown merinque in the oven.

The flavor of the honey and lemon blend well in this pie filling.

HONEY DESERTS

Its flavor and sweetness are such that honey combines well with fruits, both raw and cooked, so that it is an excellent addition to desserts.

A honey of delicate flavor, like orange, gallberry, or mangrove, should be used. It makes a delectable sweetening for whipped cream and for desserts. It supplies both sweetening and flavor for salad dressings when prepared with fruit salads. If granulated, the honey should be liquefied over hot water before it is combined with other ingredients.

HONEY TANGELO TAPIOCA

1 cup honey	2 tbsp. sugar
Pinch salt	1 cup shredded coconut
½ cup quick cooking tapioca	2 cups tangelo sections
3 cups boiling water	Whipped cream

Heat honey and water in double boiler, add pinch of salt, sugar and tapioca. Cook for 15 minutes, stirring frequently. Add shredded coconut and cook until it thickens. Cool and pour over tangelo sections, stirring lightly with a fork to mix through the tapioca. Put in refrigerator to chill very thoroughly. Serve with whipped cream or honey meringue. Sliced peaches, pineapple, mango, banana, guava, tangerine, or Temple orange sections, or a combination of fruits all provide delicious variations.

NOTE: The tangelo is a citrus fruit resulting from a cross between the tangerine and grapefruit—a combination of delightful flavor.

HONEY CUSTARD (Baked)

4 cups scalded milk	8 tbsp. strained honey
5 eggs	¼ tsp. salt
	Nutmeg

Beat eggs sufficiently to unite whites and yolks but not to make them foamy. Add other ingredients, mix thoroughly and pour into individual custard cups. Sprinkle lightly with nutmeg. Set cups in a pan of warm water, place in oven. Bake in moderate oven until when a knife is inserted into custard it comes out clean. Remove cups from water immediately. Serve hot or cold.

HONEY CUSTARD (Boiled)

2 cups milk	Salt—few grains
3 egg yolks	2½ tbsp. strained honey
	½ tsp. vanilla

Heat milk and honey in a double boiler. Beat egg yolks, add to yolks the hot milk mixture and return to boiler to finish cooking. When the mixture coats a silver spoon, remove from fire. Chill, add flavoring.

HONEY MOUSSE

¼ cup powdered sugar
½ cup shredded pineapple
    (drained)
2 egg whites
½ cup honey (warmed)
½ cup candied orange peel
    or kumquat
1 cup cream—whipped
1 tsp. vanilla extract
½ cup pecans

Mix pineapple, honey, chopped nuts, peel and flavoring. Cool. Beat the egg whites until stiff and add powdered sugar. Beat cream until fairly stiff. Fold all ingredients together and freeze either in paper mousse cups or in freezing trays of the refrigerator.

HONEY ROLL

2 cups rice or corn flakes
1 cup nuts—chopped
1 cup dates—cut in small
    pieces
1 cup honey
16 marshmallows—cut in
    small pieces

Roll flakes fine and combine carefully with other ingredients and make into a roll. Then cover with more rolled flakes and place in refrigerator until thoroughly chilled—8 to 10 hours. Serve with whipped cream sweetened and flavored with honey. Easy to make and very delicious.

HONEY BANANA MOLD

2 tbsp. gelatine
¼ cup cold water
1½ cups milk
1   lemon
½ cup honey
3 bananas (mashed through
    sieve)
1 cup whipped cream

Soak gelatine in cold water until soft. Heat milk, remove from fire and stir in gelatine. Add honey, mashed bananas, and lemon juice. Set in a cool place and when it begins to thicken fold in the whipped cream. Chill thoroughly.

HONEY ICE CREAM

One quart thin cream; ¾ cup delicately flavored honey. Mix and freeze in the usual way.

HONEY CHOCOLATE ICE CREAM

3 cups milk
3 eggs
1 qt. cream
2   squares of chocolate
⅛ tsp. salt
1½ cups mild honey

Make a boiled custard of the milk, melted chocolate, honey, eggs and a little vanilla. When cool add the cream and freeze.

FROZEN HONEY CUSTARD

4 egg yolks
2 cups water
2 cups rich milk
Pinch of salt
1 cup honey

Beat the egg yolks; add the salt and water. Cook over boiling water two minutes, stirring constantly. Cool. Add milk and honey. Freeze with 1-8 salt-iced mixture. Yield, 1¾ quarts.

HONEY GINGER SHERBET

2 quarts water
6 lemons
Cold water
½ cup preserved ginger, cut fine
3 cups honey
1 tbsp. gelatine
½ cup syrup from preserved ginger
2 egg whites

Boil water and sugar together for five minutes. Add lemon juice, gelatine softened in a little cold water, the syrup and preserved ginger. Freeze to a mush, then stir in the beaten egg whites, and continue freezing.

GRAPEFRUIT SHERBET

1 pt. boiling water
4 cups grapefruit juice
2 tsp. gelatine
2 tbsp. cold water
2 cups honey
Juice 1 lemon
Shredded or candied orange peel

Soften gelatine in cold water. Add boiling water and honey. Stir until dissolved, cool and add fruit juices. Cool and freeze in three parts of ice to one part of salt. Garnish each serving with shredded candied cherries or strips of candied orange peel.

HONEY STRAWBERRY SHERBET

1 pint strawberries
2 lemons
⅞ cup honey
2 cups water
1 egg white

Mix the strawberries (which have been put through a sieve), lemon juice, water and honey and let stand several hours to blend. Put into a freezer and when it begins to freeze add beaten egg white. Freeze with 8 parts ice to 1 part salt and pack with 3 parts ice to 1 part salt. Makes 1 quart.

HONEY PLUM PUDDING NO. 1

1 cup grated raw carrots	2 tsp. cinnamon
1 cup grated raw sweet potato	½ tsp. nutmeg
	½ tsp. allspice
½ cup chopped dates	¼ tsp. cloves
½ cup candied orange peel, citron or pineapple	½ tsp. soda
	½ cup flour
1 cup honey	1 cup raisins
¼ tsp. salt	⅔ cup suet (chopped or ground)

Combine ingredients in order given. Stir until mixture is well blended. Pour into well greased Pyrex refrigerator dish (1 qt. size) or Pyrex casserole; put cover on and bake in oven at 250° F. for 2½ hours. Remove from oven, cool without removing cover. Serve with Honey Butter.

The above plum pudding recipe is an easy one to make, is inexpensive and when served with a small topping of Honey Butter instead of the proverbial powdered sugar hard sauce, is everything taste satisfaction requires. Make up a dozen or more and use the extra ones as Christmas remembrances.

HONEY PLUM PUDDING NO. 2

½ cup oatmeal (measured after cooked)	½ tsp. soda
	2 tsp. cinnamon
½ cup allbran	1 tsp. allspice
1 cup seedless raisins	1 tsp. nutmeg
1 cup pecans	1 cup honey
¼ cup citron	1 egg
¼ cup dates	½ cup jelly (Honey guava jelly is recommended)
¾ cup flour	

Combine ingredients in order given. Bake in a covered greased pudding mold or in a covered Pyrex dish for 2½ hours at about 250° F.

HONEY CITRON STEAMED PUDDING

½ cup chopped suet
¾ cup finely sliced citron
¾ cup nut meats
½ cup honey
Juice and rind of ½ lemon

1⅝ cup flour
    Reserve ¼ cup of this flour
    for dredging
½ cup sweet milk
½ tsp. soda dissolved in a little
    hot water
½ tsp. salt

Steam 2½ hours in well greased pudding mold with horn. Steam in a deep vessel which has a tight cover and a rack in order that the water may circulate freely under mold. If necessary to add more water during steaming process, be sure water is boiling.

Remove from mold while still hot and serve with hard sauce or honey.

DATE PUDDING

¾ cup honey
2 eggs
½ cup chopped dates
½ cup chopped nut meats

1 tsp. baking powder
¼ tsp. salt
½ cup whole wheat bread
    crumbs
½ cup flour

Dust the dates and nuts with a portion of the flour. Sift the remaining flour with the salt and baking powder. Add the beaten eggs to the honey, then the crumbs, the sifted dry materials, and the dates and nuts. Mix well, pour into a greased baking dish and bake 20 minutes in a moderately hot oven.

Serve with cream hard sauce or Honey Butter.

HONEY BUTTER

2 parts honey   1 part butter

Let butter stand in room temperature until it is soft. Add honey and stir until perfectly blended. Place in glass jar which can be tightly covered and stand in refrigerator.

Uses for Honey Butter

Blend with chopped nuts as simple topping for sponge cakes.

As a service for hot biscuits, griddle cakes, waffles, instead of serving honey and butter separately.

Delicious on nut bread for tea service.

HONEY CARROT PUDDING

1 cup grated carrots	1 cup grated potatoes
1 cup raisins	1 tbsp. mixed spices
1 cup honey	1 cup flour
1 tsp. soda	1 tsp. salt
1 egg	1 cup suet

Steam for three hours. Serve with Honey Butter or Honey Kumquat Sauce.

HONEY KUMQUAT SAUCE

1 cup honey	1 cup orange juice
½ to ¾ cup finely chopped fresh kumquats, seeded	⅛ tsp. salt
	1 tbsp. butter (may be omitted)

Combine the ingredients and let stand over hot water, without cooking, for about 30 minutes to blend the flavors. Serve as a sauce on ice cream.

SALADS, SALAD DRESSING AND SANDWICHES

FROZEN FRUIT SALAD

½ cup honey meringue	2½ cups prepared fruit, pineapple, orange hearts and loquats, or guava, mango and papaya
½ cup Honey Salad dressing	

Add fruit to the salad dressing and fold in whipped cream. Turn into freezing tray of automatic refrigerator and freeze.

HONEY SALAD DRESSING

2 egg yolks	¼ cup honey
Pinch of salt	Juice of ½ lemon
½ cup cream, sweet or slightly sour	2½ tbsp. salad oil
	⅛ tsp. paprika

Beat egg yolks, then pour in the hot honey. Cook for a moment, beating continually, then fold in the salad oil, lemon juice, the cream beaten stiff and the seasonings.

HONEY MAYONNAISE

1 egg
1 tsp. salt
2 tbsp. honey
1 tsp mustard
6 tsp. honey vinegar

6 tsp. lemon juice
1½ cupfuls salad oil
Paprika
Few grains cayenne

Into a conical shaped bowl break an egg and add the salt, honey, mustard, dash paprika, the cayenne and 1 tbsp. honey vinegar. Beat thoroughly with a good egg beater and add the oil, 1 tbsp. at a time, beating thoroughly after each addition until ½ cupful is added and the dressing is thick. Then the oil can be added in larger quantities at a time. When one cupful has been added, dilute with the rest of the oil. Use altogether 1½ cupfuls of oil, beat vigorously all the time during the making. When finished, dressing should be smooth and thick.

HONEY CREAM DRESSING

2 tbsp honey

1 cup whipped cream
1 tsp. prepared mustard

Mix the mustard and honey together and stir in cup of whipped cream. Adds a piquancy to pineapple salad combinations.

HONEY CHEESE DRESSING

2 oz American cheese
2 tbsp honey

3 tbsp. whipped cream
1 cupful honey mayonnaise

Mash cheese, add whipped cream, then honey. Stir in honey mayonnaise. This dressing is nice for peas, tomatoes, or asparagus salad.

HONEY PEANUT BUTTER SPREAD

½ cupful honey

½ cupful peanut butter

Blend peanut butter and honey. More honey may be added if a sweeter paste is desired. Excellent on hot buttered toast or as a dressing for sweet sandwiches.

HONEY OATMEAL OR NUT BREAD AND CREAM CHEESE SANDWICHES

Spread thin slices of honey oatmeal or nut bread (at least three days old) with honey cream cheese paste. Place buttered slices with cheese spread slices together, cut crosswise and allow three triangles to each serving.

HONEY CREAM CHEESE PASTE

1 cake cream cheese
3 tbsp. honey
   Mix into paste

3 tbsp. chopped salted pecans

CHICKEN AND GREEN PEPPER SANDWICHES

Spread 20 thin slices of bread with butter; then on 10 of them place thin slices of white meat of cooked chicken; on other 10 spread a mixture of chopped green pepper and honey salad dressing. Place crisp white lettuce on the latter; press together with chicken covered slices, cut and serve with chilled olives and sliced tomato as garnish.

HONEY AND CREAM CHEESE SANDWICHES

Mix honey with cream cheese and use as filling for sandwiches. Chopped nuts, dried or crystallized fruit or peanut butter may be added to the cheese.

HONEY, MILK AND OTHER DRINKS

"A land flowing with milk and honey," was the description of Canaan, hence, honey and milk even in Biblical times were recognized as valuable foods. Honey sweetened fruit-ades, iced tea and coffee are given an added flavor that is very delicious. The amount to use depends on personal taste. Honey is convenient to use in hot tea, just a teaspoonful or more from the honey jar as desired, but for cold drinks the honey should be blended with a little warm water, before adding the iced beverage.

HONEY MILK SHAKE

Mix one dip of ice cream with ¼ cup honey. Add 1 cup milk and shake well in malted milk mixer.

HONEY EGG MILK SHAKE

2 eggs
Thin cream
1½ cups ice water
6 tbsp. honey
Chipped ice

Beat eggs well and pour into glass fruit jar or shaker. Add remaining ingredients and shake. Yields, 3 servings. May top each glass with whipped cream.

HONEY ORANGE COCKTAIL

Mix juice of 6 oranges, 6 tbsp. honey and few grains of salt. When ready to serve, shake up with ice cubes and add shreds of yellow orange rind. Decorate with sprig of mint.

VITALITY COCKTAIL

Juice of two oranges, juice of ½ lemon, yolk of 1 egg, warm honey. Beat the ingredients together and drink every morning.

HONEY COCOA

4 tbsp cocoa
2 to 4 tbsp. honey
Dash of salt
1 cup cold water
3 cups milk

Mix cocoa, sugar, salt, and water in upper part of double boiler and place over direct heat. Stir until smooth; boil 2 minutes. Place over hot water, add milk and heat. Beat well, using rotary egg beater, and serve at once.

HONEY ICED CHOCOLATE

Blend 2 tsp. cocoa with 3 tsp. honey. Let 1 cup milk come to boiling point. Remove scalded milk from fire, add honey and cocoa mixture and pinch of salt. Stir well. Pour this mixture in iced tea glass filled with cracked ice. Top with whipped cream. For hot chocolate, omit ice and add ¼ cupful of scalded milk.

HOLLYWOOD HONEY PUNCH

Juice 12 lemons
Juice 12 oranges
3 quarts water
1 pt. tamarind juice
1 pt. guava juice
1 pt shredded pineapple
Honey to sweeten

Warm honey and add to water. Blend and add fruit juices and shredded pineapple and chill. When ready to serve, garnish with thin slices of lemon and orange and pour over ice.

GUAVA HONEY PUNCH

1 cup honey	2 cups water
2 cups pared and seeded guavas and juice	¼ cup lemon juice
	½ cup orange juice
	Mineral or ice water

Simmer the honey and water together until blended, set aside to cool. Force the guavas through fruit press and combine the pulp with the orange and lemon juice. Add to the cold syrup and let chill thoroughly. Just before serving, strain and dilute to taste with mineral or ice water. Peaches, plums, mangoes, may be used the same way as guavas.

TROPICAL COOLER

Guava juice	Cracked ice
Juice fresh limes, calamondins or tangelos	Honey to taste

Blend well and serve with thin slices of fruit.

HONEY ICE CREAM SUNDAE

Over a serving of ice cream—usually vanilla or chocolate is preferred—pour a generous stream of gallberry, orange or mangrove honey.

HONEY WITH VEGETABLES AND MEATS

Carrots, green or wax beans, beets, squash, turnips, sweet potatoes, and other vegetables—important in the diet—are better flavored through the addition of a small amount of honey.

Use a teaspoonful of mild honey to each cup of vegetables when adding other seasonings.

HONEY BUTTERED BEETS

2½ cups beets, cooked and sliced
½ cup boiling water
1 tbsp. flour
4 tbsp. vinegar or lemon juice
2 tbsp. butter
4 tbsp. honey

Blend butter and flour, add hot water and stir until smooth. Add other ingredients and pour over the beets that have been placed in a buttered baking dish. Cook 20 minutes in moderate oven.

HONEY CUSHAW EN CASSEROLE

2 cups cushaw, pared and thinly sliced
1 cup apples pared and thinly sliced
3 tbsp. butter
1 tsp. salt
½ to ¾ cup honey, warmed

Place a layer of cushaw in buttered baking dish, then a layer of sliced apples. Add salt, dot with butter, and cover with honey. Add another layer of cushaw and apples and dress as before with seasonings.

Top with a layer of cushaw, brush with butter and bake in a hot oven for 45 minutes, covering the first half of the time. Sweet potatoes may be baked in the same way as cushaw or other winter squash.

Delicious to serve with broiled chicken or honeyed ham.

BAKED SWEET POTATO WITH HONEY AND MARSHMALLOW

Scrub sweet potatoes as for ordinary baking. Bake until soft. Then with sharp knife cut across on top. In this opening drop first 1 tsp. honey, then press in half a marshmallow. Return to oven and heat for just a few seconds. The honey is absorbed almost immediately by the hot sweet potato and marshmallow is toasted just enough by the few seconds of reheating. Serve at once. Especially nice for crown roast of pork, roast chicken, or turkey.

FRIED SWEET POTATOES

Boil 10 medium large sweet potatoes with skins on. When about two-thirds cooked, remove from fire, run cold water over them. Pare, slice in ⅛-inch slices and put in frying pan well greased. Fry until browned, then add a mixture of ½ cup honey and ¼ cup brown sugar. Stir through sweet potatoes, let remain over low flame for three minutes. Serve at once. (27 servings.)

HONEY WAY CHILI

1 cup diced celery
1 cup chopped onions
4 cups ground beef
1 tsp. chili powder
1 pt. tomato puree

1 qt. red beans (cooked or canned)
1 pt. water
1 tbsp. salt
6 tbsp. honey

Fry beef, onions, and celery slowly for about one hour. Should be thoroughly browned—being careful not to burn onions or celery. Place one quart of red beans either cooked or canned, 1 pint tomato puree, 1 pint water, and 1 tbsp. salt in kettle. Let come to a boil, then add fried meat and vegetables. Simmer slowly for two hours. Then add chili powder, and just before serving stir in honey. Serve piping hot.

FESTIVE HONEYED HAM

For a delicious ham which requires a minimum amount of holiday preparation, the ham should be given its preliminary cooking the day before. The whole or half ham is brought to a boil, then simmered, allowing 20 minutes to the pound. Use from 1 to 2 cups of pineapple juice in the water in which the ham is boiled.

Remove ham from liquid, skim and pour over the skinned ham 2 cups of honey (for ham weighing 9 to 10 lbs.). Let stand over night. In the morning add enough liquid which has been reserved from the boiling liquor for basting purposes. Rub the skinned surface with bread crumbs, then baste frequently with honey liquid to which has been added a cup of raisins or 1 cup spiced roselle.

HONEYED HAM AND PINEAPPLE

Brown a rather thick slice of cured ham in a baking dish, pour 4 tbsp. honey over ham and stick 3 or 4 cloves in the ham. Place pineapple rings on ham and bake in moderate oven, covered for the first 10 minutes. In place of pineapple, apples, sweet potatoes, or carrots may be used and pork chops may be substituted for the cured ham.

BAKED APPLES WITH HONEY

Bake apples with a bit of water until tender. Butter may be added if desired. Remove from oven, drizzle honey over hot apples. The hot apples will readily absorb the honey and by the time of serving, the honey will have permeated the apple tissue and blended to form a perfectly delightful dish.

HONEY BAKED HAM

1 lean ham (weighing from 7 to 9 lbs.)
15 cloves
Celery leaves from one bunch of celery
½ cup honey
2 eggs, beaten
1 tsp. cinnamon
1 qt. honey vinegar or pickle juice
Honey raisin sauce
Soda
Boiling water

Thoroughly wash the ham, rub soda over the surface; rinse in cold water. Celery leaves, cloves, cinnamon, honey vinegar and ¼ cup honey should be placed in a kettle full of boiling water. In this place the ham and simmer until perfectly tender —about five hours. Remove the skin after taking ham from kettle, and brush with beaten egg and honey (2 eggs beaten blended with ¼ cup honey). Stick in about 30 cloves at even intervals and brown in very hot oven.

Serve with Honey Raisin Sauce.

HONEY RASIN SAUCE

1 cup raisins
1 cup honey
½ cup water

Cook very slowly until raisins are soft but not mushy. Add honey and a teaspoonful of lemon juice and serve over ham slices.

HONEY CONFECTIONS

Home made candies are always a special treat, but when honey is used in their making, they are doubly delicious. In candy making, honey imparts its own individuality to the product and opens up a wide range of interesting opportunities in the candy way.

HONEY FUDGE

2 cups white sugar
1 cup milk
¼ cup honey
2 inch square chocolate
1 tsp. vanilla

Allow to cook to soft ball stage. Cool. Beat 20 minutes after cool.

HONEY CARAMELS

2 cups granulated sugar
2 cups honey
½ cup butter
2 cups rich milk
1 tsp. vanilla

Choose a heavy iron, aluminum or copper kettle for cooking. Stirring occasionally, boil sugar, salt and honey to 245° F. Add butter and milk gradually, so that the mixture does not stop at any time. Cook rapidly to firm ball stage (256° F.) Stir constantly because the mixture becomes very thick and sticks easily at the last. Add vanilla and pour into a buttered pan.

Cool thoroughly before cutting. Cut with a heavy sharp knife, using a saw-like motion. Yield, 2 lbs. or 45 caramels ¾ x 1½ inches.

HONEY PEANUT BRITTLE

2 cups sugar
1 cup honey
1 tbsp. butter
1 cup water
⅛ tsp. salt
2 cups roasted peanuts

Put sugar, honey, salt and water in saucepan. Stir until sugar is dissolved. Cook to 300° F. Remove from fire. Add butter and peanuts. Stir just enough to mix thoroughly. Pour out on a well greased marble slab or baking sheet into very thin sheets. Allow to cool and break into irregular pieces.

HONEY BUTTER SCOTCH

2 cups honey
2 cups sugar
1 cup butter
1 tbsp. cinnamon

Boil ten minutes or to crack stage, 290° F., and then pour into a buttered pan and when cold cut in squares.

HONEY DIVINITY

2⅓ cups sugar
½ cup honey
¼ tsp. vanilla
2 egg whites

¼ tsp. salt
1 cup water
¾ cup coconut or nut meats

Put sugar, honey, salt and water into a saucepan and cook, stirring until the sugar is completely dissolved. Continue cooking, without stirring, until a firm ball is formed in cold water, or until 268° F. is reached. Wash down any sugar crystals that may form. Remove from fire and slowly pour the syrup over the egg whites which have been beaten until stiff during the latter part of the cooking of the syrup. Beat during this addition. Continue beating until the candy will hold its shape when dropped from the spoon. Add vanilla and nuts or coconut; mix thoroughly. Drop from teaspoon onto waxed paper.

If taken off when temperature of 262° F. has been reached, it can be used for the following:

Stuffing dates—Making coconut balls—Shaping in balls and dipping in chocolate.

This may be varied by the addition of candied fruits or nuts. These chocolates thus made are delicious.

ORANGE BLOSSOM TAFFY

2 cups orange blossom honey
1 cup boiling water

2 cups sugar
1 tsp. vanilla

Put honey, sugar and water into sauce pan; stir until sugar is well dissolved. Place on fire and cook to 270° F. Remove from fire; add vanilla. Pour out on a well-buttered dish. When cool enough to handle, pull until creamy and stiff like other taffies.

HONEY TAFFY

2 cups sugar
½ cup strained honey

⅔ cup water
1 tsp vanilla

Put all of the ingredients except the vanilla into a sauce pan and cook, stirring only until sugar is dissolved. Continue cooking until a hard ball forms in cold water or the temperature 263° F. is reached. Remove from fire and pour into buttered pan. When cool enough to handle, pour vanilla into center of the mass, gather the corners and remove from the pan and pull.

When candy is white and rather firm, stretch out into a long rope and cut into pieces of desired size, using scissors for the cutting. Nut meats may be added just before the taffy is ready to cut, which must be worked in during the pulling.

HONEY ORANGE STRIPS

Remove the peel from 3 oranges in quarter sections, then cut into strips with scissors. Cover the rind with salt water in the proportion of 1 tbsp. of salt to 1 quart of water and let stand over night. Drain and cover with cold water, then bring to the boiling point; repeat this process three times. Then if tender, rinse in cold water, drain, then simmer very slowly in 1 cup of honey from 45 to 60 minutes. Remove the rind with a fork, drain and lay on waxed paper. Allow to dry for a day or two. The strips may then be coated with chocolate, if desired.

Grapefruit may be prepared in a similar way but grate rind carefully before cooking tender in an abundance of water. Drain, then cook the peel in a syrup made with 2 cupfuls of honey, 2 tbsp. lemon juice or grapefruit juice.

Cook the grapefruit strips one hour or more, then allow them to stand all night in the honey syrup. Remove with a fork and lay on waxed paper for a day or two. These may be coated with milk chocolate or bitter chocolate.

HONEY SPECIALTIES

HONEY MERINGUE (Uncooked)

Electrical Beater: Use one egg white to one-half cup honey, placing in bowl or electrical mixer and turning to speed 2, allowing mixture to whip until it peaks.

Hand Beating: Place one-fourth or one-third cup honey in bowl with one egg white and beat with double Dover or Ladd improved (ball bearing type) beater until stiff.

This mixture keeps indefinitely when kept uncovered in refrigerator. Honey meringue made with granulated honey keeps just as well and in some cases has been found to whip up more easily by hand than when strained honey is used.

Honey meringue may be used as a topping just as whipped cream or marshmallow is used, on top of pie; for toasting as ordinary meringue; on sweet potatoes; mix with rice crispies and use as a paste to spread on butter wafers for tea; as a dressing for fruit salad; delightful for date tortes. The amount of honey used depends entirely upon the individual preference for the honey flavor.

Add 2 tbsp. melted butter to 1 cup meringue for a good gingerbread topping.

HONEY TOAST

Trim slices of bread (slices should be about 3/8-inch thick). Toast properly, then butter and brush with honey. Reheat enough to have toast absorb honey and serve piping hot.

HONEY CINNAMON TOAST

Spread slices of fresh toast with butter, brush with honey (about 1 tbsp. honey for each slice), sprinkle with cinnamon and oven toast enough to blend cinnamon and honey.

HONEY NUT BREAD TOAST

Place thin slices of honey nut brown bread on thin pan, oven toast both sides, spread with butter and honey. Cut in triangles and serve open face.

These breads must be oven toasted and very carefully turned over on flat tin with spatula so that the slices will remain intact. Hot honey nut bread is delicious when spread with orange marmalade immediately when removed from oven.

Any of these toasts must be served piping hot to be good.

MISCELLANEOUS

HOT HONEY LEMONADE

Hot honey lemonade is particularly valuable in relieving the grippe. When suffering from a cold, take a hot honey lemonade just before retiring.

Four tbsp. lemon juice mixed with 4 tbsp. honey. Add 1 cup boiling water. Drink hot.

CANNING AND PRESERVING

Honey may be substituted for part, or in many cases where fruits are of high flavor, for all of the sugar needed in canning and making jelly, jam, preserves, fruit pickles and conserves. Of course, where all honey is used it tends to mask the more delicate flavor of the fruits, and color and texture of the product too is darkened somewhat. It is necessary, therefore, to use the mildest flavored honeys in order that the individual, distinctive fruity flavors may not be too much overshadowed by that of the honey.

Flavors of honey also vary with age and storage, so it is always desirable to use a new honey for canning purposes when available. The honey flavor combines better with some fruits than with others. A combination of fruits for making conserves or jams and butters in which spices are used, for instance, is usually more pleasing than that made with one fruit alone.

In using the honey, two precautions should be observed:

1. Since honey has a tendency to foam considerably when heated, there is some danger of the products "cooking over" at the beginning of the cooking period, if not watched carefully.

2. Since honey is part water, in order to obtain the desired consistency, it is necessary to cook the product in which it is used slightly longer.

Basic recipes for honey syrups call for either an all-honey syrup, one cup honey and three cups water, or preferably a honey and sugar syrup, ½ cup sugar and 3 cups water. This syrup is recommended for use with mild flavored fruits, like figs, grapes, loquats, mangoes, peaches, pears, pineapple, cultivated plums.

For fruits with more tartness, like the sour guavas and many of the wild plums, a heavier syrup may be desirable. The amount of dilution required for the syrup will vary with the quality of the honey and the degree of sweetness preferred. Use less rather than more sweeting. When all honey is used sometimes lemon juice is added to the all-honey syrups to counteract sweetness and to give an interesting blend of flavors.

To prepare an all-honey syrup, bring water to boil, add honey, let boil again, skim and strain and it is ready for use. For the honey and sugar syrup, bring water and sugar to boiling point, add honey, let boil again, strain and use. Prepare fruit or berries, pack into hot containers, add hot syrup in the same way as a sugar syrup; seal and process product according to the standard time table for canned fruits. Berry juice, grapejuice or other fruit juice may be used to advantage in place of water for these canning syrups.

HONEY JELLIES

Jelly of one's favorite honey is easily made when the required pectin, liquid fruit pectin, or a powdered citrus pectin, is provided. Honey jelly takes very little time and makes a clear, amber product of the pronounced flavor of the honey used. In making jelly without using an added pectin, strong flavored juices, high in both pectin and acid, are essential or a jelly of a gummy texture will result. Crab-apple, mayhaw, wild plum, sour guava juices that give a high pectin test, are good jelly juices to use with honey, particularly when ½ the honey is replaced with sugar for the fruit juice combination.

HONEY JELLY

2 cups honey   ¼ cup liquid pectin or
½ cup water   1 teaspoon powdered lemon pectin*

Combine water with honey and heat very gently to avoid scorching and the development of off flavor. Stir constantly until boiling, then add ¼ cup liquid fruit pectin, bring just to boiling and immediately remove from heat. Pour into hot, sterilized glasses. The yield will be about 4 small glasses.

LEMON HONEY JELLY

¾ cup lemon juice   2¼ cups mild flavored
½ cup liquid fruit pectin†   honey

Combine lemon juice and honey. Bring carefully to a full rolling boil. Add liquid pectin, stir constantly and bring just to a boil. Pour into small glasses and seal.

CRAB-APPLE HONEY JELLY

2 cups fruit juice   ¾ cup honey
2 tablespoons lemon juice   ¾ cup sugar

Mix juices and boil 5 minutes. Add sugar and bring to boiling point. Add honey and cook to jelly test (220° F.) or until the jelly stage is reached, as indicated by the flaking or sheeting from inside of spoon. Make pectin test before starting jelly making preparations to be sure a good textured jelly can be made from the fruit juice on hand. Guava juice, mayhaw or other juice high in pectin may be used the same way as crab-apple juice.

---

*When powdered lemon pectin is used, heat the honey gently to about 155° F. In another pan or kettle heat the water to about simmering. Remove a small part of the warmed honey into a cup and stir the dry pectin into it, making a smooth paste. When the pectin and honey are well mixed pour into the hot water. Rinse the cup with the pectin solution until all the mixture has been transferred to the water solution. Stir and heat until the pectin is completely dissolved. Be sure there are no lumps remaining. Add the pectin solution at once to the honey, which should be about 155° F. Bring to a temperature of about 170° F. or slightly higher. Pour into small containers and seal at once. The high temperature required in the usual jelly making procedures should be avoided in making honey jelly, particularly when powdered lemon pectin is used, as toughness, a darker color and a rather strong flavor would result after a few weeks of storage.

†Manufacturers of citrus fruit pectin in California and of apple pectin in New York, Missouri, and elsewhere, furnish full and detailed directions for the use of their respective products. If pectin (commercially speaking) is used, in connection with the fruit juice, it must be declared on the label.

SOUR ORANGE PRESERVES

The fruit of the native sour orange, so generally used for root stock over many portions of the citrus area, is used for making delightful preserves that are always popular.

For best flavor use the fruit when well matured and highly colored. Grate off all oil cells leaving the rich yellow colored skin exposed. Cut into quarters and remove from pulp. Soak the peel in salt water (1 cup salt to 1 gallon water) overnight. Squeeze juice from pulp and save to add to preserves during the last cook. Drain peel from salt water. Cover well with clear water and boil for 10 minutes. Drain and cover with fresh water and cook until peel is tender. If no bitter flavor is desired, it may be necessary to change the water several times. However, if the fruit used is fully ripe the slightly bitter flavor is agreeable to most palates.

Drain peel and drop into a hot syrup made of three cups honey and two cups water for each 2 pounds of peel. Cook until peel is clear and syrup somewhat thickened. Remove from heat and let stand overnight. The next day, take from syrup, add ¾ cup honey and ½ cup sour orange juice and bring to boil. After boiling 10 minutes or until thickened, replace fruit. Boil another 10 minutes or until syrup is thick. Pack into hot jars immediately and process pints for 10 minutes at boiling. Grapefruit, tangelo and shaddock peel may be preserved in the same manner as the sour orange.

HONEY KUMQUAT PRESERVES

Clean kumquats and puncture carefully. Drop into slightly salted water and soak overnight; next day pour off salted water, cover well with fresh and bring to a boil. Drain and cover again with fresh water and cook until tender. Drain.

To one pint of fruit add ½ pint of sugar, ¼ pint orange honey and one pint of water or orange juice. Drop fruit in the boiling syrup and simmer until clear and syrup is slightly thickened. Plump overnight in the same vessel, covering tightly while still boiling and removing from fire. The second or third day place back on the fire and cook until syrup is heavy. Pack in jars as any preserve, or if candied kumquats are desired for immediate consumption, drain, put on wire rack to dry and, while still sticky, roll in granulated sugar.

SWEET FRUIT PICKLES

(Peach, Pear, Pineapple)

2 cups mild flavored honey
1 cup cider vinegar
1 cup water
½ piece ginger root*
Fruit

½ lemon sliced, or 3 calamondins or kumquats, cut in thick slices and seeded
3 inches stick cinnamon
12 whole cloves

Combine honey, vinegar, and citrus fruit and spices. Heat to boiling and boil gently about 5 minutes. Have ready 4 to 6 cups of the quartered pears, peach halves or pineapple chunks. Add to spiced solution. Cook until just tender. Pack fruit in hot jars, cover with the boiling syrup and seal at once.

PEAR AND GINGER CONSERVE

½ lb. green ginger scraped and chopped
6 lbs. honey
8 lbs. pears weighed after paring and coring

1 pint water
4 oranges
3 lemons, juice and thinly shredded peel
2 cups pecans or black walnut meats

Cook the ginger, orange and lemon peel with a pint of water until tender, then add honey, orange and lemon juice; cook, put in the pears chopped coarsely and cook until pears are tender. Add nut meats. Cook five minutes longer. Pour in small hot jars and seal, boiling hot.

---

*Ginger Zinzibar, officinale, often confused with the common ornamental ginger lily, grows well in Florida and produces choice roots if given rich soil, sufficient moisture and semi-shade. It is an erect herb, 12 to 24 inches high, canna like in appearance and grows from thickened rhizones which branch fingerlike and send up new shoots from the tips near the surface of the soil. If desired for preserving and candying, the roots should be dug while tender and succulent, rather than when old, tough and fibrous. Ginger is one of the world's most popular spices. It is an indispensable part of chutneys, giving them much of their spiciness and pungent flavor.

Honey Bees and Their Products

By T. J. BROOKS
(Late Assistant Commissioner of Agriculture)

It would seem that as old a subject as Honey Bees and their products would have long since been exhausted and nothing new could be said on the theme. But it seems that no subject is really ever "worn out" as we never know all about anything. The bee industry has been revolutionized during the last fifty years.

Honey is the oldest of all the sweets used by man. There seems to be no country that can claim to be original home of the honey bee. Different species were found in practically all the inhabitable parts of the world. The aborigines of Peru sacrificed honey to the sun. Stingless honey bees of Brazil produced every variety of honey from good edible kind to black and sour. No one knows who first tasted honey and pronounced it good. Samson, the strong man, made a riddle on honey he found in the skull of a lion which he had slain. That riddle got him in trouble. John the Baptist's food, we are told, was locusts and wild honey.

The honey bee is quite a useful animal. He does no damage to the plant from which he gets his product—he is beneficial in his visits to flowers by carrying pollen and aiding in fertilization of the seed germs—and he brings a valuable product to the service of man. He is one creature that seems to be miserable unless he is at work. His industry is his life.

The U. S. Department of Agriculture reports that the average analysis of honey shows the following percentages of elements:

Water	17.7
Laevulose	40.5
Dextrose	34.02
Sucrose	1.9
Dextrin and gums	1.51
Ash	.15
	95.78%

This leaves 4.22 percent unaccounted for. These percentages differ largely in different specimens analyzed. Extraneous matter gets into some honeys, such as pollen or peculiar substances that may be in the nectar as extracted from the flower.

All edible honeys are thought to contain vitamins A, B and C—neither of which can be found in cane or beet sugars, according to authorities on the subject. The proportions of laevulose and dextrose vary greatly in different flowers from which honey is obtained. A high percentage of laevulose prevents crystallization. The tupelo of the southeastern states and the sages of California produce this kind of honey. The high percentage of dextrose causes honey to crystallize quickly and is therefore less desirable for keeping indefinitely and for shipping long distances.

It remains for the physicians and dietitions of this generation to discriminate between the different sweets used for food and classify them according to their food values and dietetic qualities. Even honeys are not all alike in content, flavor or appearance. The world today is so completely commercialized that one may look for a flare-up if he says that one kind of sweet is better for the human anatomy than another.

The general keeping of bees is a good thing economically, in spreading pollen and in furnishing honey for the household. But the fact remains that the production of a certain variety and quality in large amounts is the only way to open up a sure market at a good price. Buyers of large quantities of anything want to know if they can depend on the source of supply to be ready when they want it and in the quantities they want. This is the only way they can build up a trade that continues from year to year. The human taste is subject to cultivation and when customers of dealers in honey ask for a certain honey or syrup they have cultivated their taste to that particular kind and do not want to be put off with "something just as good." If the orange honey producers were to advertise their honey through some central office it would vastly increase the market. The same is true of the tupelo honey or any other good variety. Melilotus honey is of a kind and appearance that appeals to hundreds of thousands but it takes advertising to create and hold buyers.

The State Department of Agriculture has nothing to do with the supervising or inspection of bees or honey. That comes under the jurisdiction of the State Plant Board. As the extermination of plant pests is a Plant Board function it has been construed that bee pests should come under the same head.

I have no comparative figures of the value of honey and molasses but the time was at the turn of the century when honey exceeded in value the molasses in the United States. Modern methods of refining and advertising artificial sweets have placed them far in the lead as food products.

Many physicians and dietitions are recommending honey for arthritis and neuritis. It has proven to be efficacious in many cases where all other remedies had failed.

There should be established a clinic in some institution equipped for scientific experimentation as to the value of different kinds of honey both as a food and as a remedial agent for human ills.

It would be a signal service to humanity if some medical school or hospital would establish with certainty the facts connected with this subject.

Florida is a honey-producing state, largely because we have an abundance of different nectar-producing flowers and also because of the long season during which honey can be gathered. I am of the opinion that the greatest thing the honey-producers could do for their marketing advantage would be to organize and place a fund for the judicious advertising of the distinct types, giving emphasis to the distinguishing qualities of each.

THE MYSTERY OF SWEETS

The word sweet has a multiplicity of meanings.

There are sixty English words that begin with "sweet," and as many that begin with honey.

It applies to taste, smell, looks, acts, characteristics—if pleasant. Sucrose, Dextrose, Lactose, Maltose, sacharose, levulose, glucose, are sweets. The last named is often given directly into the bloodstream.

The antonyms are sour, bitter, offensive, ugly, contemptible.

There is a universal demand for sweets. Those of taste call for sugar, honey, syrup of varying kinds and flavor. The oldest sweet known is honey. Man cannot manufacture it.

Nature has provided a little worker in the form of a honey-producing insect which gathers nectar from flowers that furnish this particular form of sweet.

An important thing to be remembered is that no other sweet has the food value that honey has. Why do you men who are in the honey-producing business not emphasize this and advertise it to the public. Are you afraid that you will make a claim that you cannot substantiate? Well I know of no more dependable authority on scientific questions than the Encyclopedia Britannica. Listen to what it says:

"In most countries at present, the amount of cane and beet sugar exceeds the honey used by fifty times, whereas in ancient times honey was the most important source of sweetness. There is, of course, much evidence that the present excessive use of artificially manufactured sugars and syrups is DETRIMENTAL. All such sugars and syrups are wholly DEFICIENT in vitamins and have had EXTRACTED from them MANY OTHER FOOD CONSTITUENTS in the manufacturing processes; just as occurs in the highly developed manufacture of other modern food-stuffs. The recent protest against artificially manufactured foods is resulting in an increase in the advice that honey be used as a natural food product, in place of such large quantities of manufactured sweets. Various new and important uses are now being found for honey, in which other syrups cannot be employed satisfactorily."

The Encyclopedia Americana has this to say:

"Honey is highly nutritive, especially as a fuel for energizing the body, as four-fifths of its components are carbohydrates. It has well recognized medicinal properties."

It is a subject of common discussion that white bleached, starchy flour makes non-nutritious bread. It is also well established that white, refined sugar is not the food that unbleached sugar is. Another thing known is that all syrups partake of the soils from which they grow and if the soil is deficient in the minerals that food should have the syrup is also deficient. What other foods than honey have a "hundred uses?"

This is true of all crops. So much so it is with honey that the same flower will produce honey in some states and will not in others. Take alfalfa; it will make good honey in the irrigated West and will not in the section east of the Missouri River. Buckwheat will produce honey in some states and not in others.

Pity that all food products could not be labeled with the statement giving the mineral content of the soils from which it grew. Big canning concerns that have their fruits and vegetables grown under contract could begin the practice and make the custom almost universal. Then the buying public would know whether they were getting a balanced food or not.

If honey is a more wholesome and nutritious food than any other sweet food the public is due to know it. As yet, so far as I know, the honey producers have never had a nation-wide advertising campaign. Why not quote the thing I have just quoted from the Encyclopedia and paste a label on each container giving the buyer the advantage of the information? That would be just common business practice and be perfectly honorable.

All honeys are not alike. There are black honeys in South America that are poisonous. There are honeys that are mixed with bitter elements. There are honeys from noxious weeds, from grass, from trees, from shrubs and even from leaves where certain creatures have left a deposit. Most honey will finally turn to sugar. That which comes from the Tupelo tree blossoms will not because it is low in dextrose and has plenty of levulose.

Glucose can be introduced directly into the blood stream. It is one of the sweets. When eaten the sugars are quickest to furnish nourishment of all foods. The mineral contents of honey depend on the flower from which it came and the contents of the flower are determined by the soil from which it grew. Of course one flower will obtain its nectar from the soil and another flower will get a different assortment from the same soil.

You have a honey in this part of the country that has levulose but little dextrose. For that reason it never crystallizes. Doctors prescribe it in some places for diabetes, arthritis, neuritis, etc. If the people generally were convinced that honey was the healthiest sweet possible there would be no surplus on the market. You should advertise as others do. Why not cultivate a honey appetite?

The honey bee is a remarkably useful animal. He not only collects a splendid article of diet but he also benefits the crops from which he secures his honey. The pollen that sticks to him as he crawls in and out of each bloom helps to fertilize the flower that it may bear fruit. It is necessary that there be a mixing of the male and female parts of different flowers for there to be a full crop of fruit. The bee also builds a sanitary container of wax for his honey from the materials that result from his work in gathering his sweets. Citrus fruit is very dependent on this pollenization process. Other crops are largely dependent on insect pollenization.

In years past, Florida was a "happy hunting ground" for beekeepers from other states. There was no prejudice against them until they began to abuse the privilege and brought foul brood into the state. It is hard enough to keep this bee pest down when the best precautions are used, but when no regard for the welfare of our honey producers is manifest it naturally brings resentment. The legislature of 1947 passed amendments to our law on apiaries prohibiting the importation of hives into the state and offered the only safeguard for our home beekeepers.

There are more than ten thousand beekeepers in Florida and they have some 238,000 colonies.

Honey and Nutrition

Every person is provided with a canal, the inside of which is provided with absorbent ducts called villi, that extract the nutritive elements from the food and pour them into the bloodstream. The blood carries these elements to the millions of cells in the body. Different parts of the body demand different elements which can be had only if the bloodstream has them: The bones need different minerals from fats, the nerves different from the glands, the all different from the brain, etc.

Now if the minerals needed are not to be had at any one point, what happens? It is "passed up" and the bloodstream flows on. What becomes of a cell if it is continually passed up? It starves! Suppose it happens that the deficient element applies to the cells of the brain! Well, why are there so many in hospitals with mental disorders? Ill health can cause worries, troubles, anxieties, despondency, forebodings and ailments galore. When we get old and begin to "slip" it might be lack of brain cell nourishment. Heart trouble! Yet, we speak of a person having a "Heart Attack" eh! A heart does not attack. It succumbs to overwork because it is weak and cannot stand the strain that it could if properly nourished. Food that produces muscle is needed. The heart is a faithful muscle. It works whether we are awake or asleep. If it stops to rest—goodbye.

People may eat plenty of good food, well cooked and still starve—literally starve, and never know it. So much greater variety of food elements are needed than is obtainable by regular channels. The fifth biggest business in the United States is canning foods. Most of the materials canned are contracted for by the canners before they are planted. Some day a farseeing canner will see his opportunity and have it in his contracts that the grower must mineralize his soils as per directions and require at least a dozen minerals in certain proportions for each crop—according to the soil's original content.

When these crops are canned a label will be placed on each container with the guarantee that "THE CONTENTS OF THIS PACKAGE WERE GROWN ON SOILS MINERALIZED AS FOLLOWS"—followed by a list of the minerals.

Believe it or not the housewife and the restaurant man will fall for this and try it at any reasonable cost. Other canners will be forced to do likewise. The result will be better health for the country and longer life.

There is more interest being taken in these problems than ever before. I find the greatest demand for bulletins on these subjects comes from educators, dietitians, physicians and experiment station operators and journalists. Honey should feature largely in all these discussions.

HONEY CERTIFICATION LAW
Florida Statutes Chapter 586

586.01 Short title.—

This chapter shall be known as the Florida honey certification law.

586.02 Definitions.—

As used in this chapter:

(1) The term "department" shall mean the department of agriculture of the State of Florida.

(2) The term "commissioner" shall mean the commissioner of agriculture of the State of Florida.

(3) The term "certified honey" shall include honey which is principally of one type or variety, such as tupelo, orange blossom, saw palmetto, gallberry or mango as shall have been inspected during its period of production, extraction and preparation for market by the department or its authorized agents and found to be reasonably free from a mixture of other types or varieties of honey, and meet other requirements as specified in the rules and regulations issued by the commissioner under the provisions of this chapter.

586.03 Inspection and certification of honey.—

(1) Any producer of honey located in Florida may make application to the commissioner for inspection and certification of his honey crop under such rules and regulations as the commissioner may issue.

(2) The commissioner, or his authorized agents, shall issue such certificates of inspection and designate or provide such official tags or labels for marking containers of "certified tupelo honey," "certified orange blossom honey" or certified honey of other identifiable types or varieties, and establish such standards of grade and quality, as are necessary to safeguard the privileges and service provided for in this chapter.

586.04 Fees for certification.—

The commissioner may fix, assess and collect, or cause to be collected, fees for the certification inspection service, the same to be paid in such manner as he may direct. Such fees shall be large enough to meet the reasonable expenses incurred by the commissioner or his agents in making such inspection as may be necessary for certification.

586.05 Unlawful to use words "certified," "registered," or "inspected."—

It is unlawful to use the terms "certified," "registered" or "inspected," or any form or modification of such terms which tends to convey to the purchaser of such honey that the same has been certified, on tags, labels or containers, either orally or in writing, or in advertising material intended to promote the sale of honey, except when such honey shall have been inspected and certified to under the provisions of this chapter by the commissioner of agriculture or by his authorized agents.

586.06 Rules and regulations.—

The commissioner may make all necessary rules and regulations to carry out the provisions of this chapter.

586.07 Employees.—

The commissioner may employ such assistants, inspectors, specialists and others as may be necessary to carry out the provisions of this chapter to fix their salaries and to pay same from such funds as may be available for the purpose.

586.08 Penalty.—

Any person, copartnership, association or corporation, and any officer, agent, servant or employee, thereof, violating any of the provisions of this act shall be deemed guilty of a misdemeanor, and on conviction, shall be punished by fine not exceeding one hundred dollars for each separate offense. Each fifty-five gallon drum of honey, or its equivalent in smaller containers, falsely tagged, labeled or otherwise falsely designated in contravention of this act shall constitute a separate offense.

586.09 Enforcement of chapter.—

The commissioner is vested with power and authority to enforce the provisions of this act and the rules and regulations made pursuant thereto by writ of injunction in the proper court as well as by criminal proceedings. It shall be the duty of the attorney general, the state attorneys, prosecuting attorneys, county solicitors, and all public prosecutors in each county to represent the commissioner when called upon to do so. The commissioner in the discharge of his duties and in the enforcement of the powers herein delegated may send for books and papers, administer oaths and hear witnesses, and to that end it is made the duty of the various sheriffs throughout the state to serve all summons and other papers upon request of said commissioner.

TUPELO HONEY

By J. A. WHITFIELD

It is not difficult to speak of Nature to a friendly and understanding audience. Both you and I turn to her in her primitive glory, when we seek rest, inspiration and strength to carry on. Perhaps we have been cruising on the Gulf and suddenly became filled with the urge for "fresh woods and pastures new." If we set our course up the historic Apalachicola River to the Chipola and the famous Dead Lakes and feasted our eyes on the inspiring verdure of virgin forest; if, in desire to prolong the vision, we had shut off the motor and held momentarily to some overhanging bough, our ears would have joined our eyes in estatic appreciation. The busy hum of myriad bees would reach us, sooth us and comfort us. An upward glance would disclose above and all around us thousands of fuzzy blooms, giving of their sweetness to the greatest workers in the world. You would have chanced upon industry in its pristine glory,—TUPELO HONEY TIME—latter part of April.

Most of this beekeeping country is as wild as in the days of the Conquistadores. If we pursue our investigation further, we find that the only evidences of man or civilization are the apiaries, elevated upon high platforms up and down the banks of the river. These are from five to twenty-five feet in height, from fifteen to twenty-five feet wide and from three to five hundred feet long. The hives are placed upon either side of the platform with the bee entrances pointing outward, leaving a walkway of between six and eight feet between the hives.

Aside from its mild and delicious flavor, this Tupelo Honey has distinct and peculiar characteristics that make it a preeminent product in certain fields. By analysis, it contains about twice as much levulose as dextrose, or a proportion of 23% dextrose, 46% levulose with the usual four or five percent of sucrose. The average American honey contains about 39% of levulose and 34% of dextrose. The higher percentage of levulose in Tupelo Honey makes it a product that DOES NOT GRANULATE. Samples have been kept for twenty-five years without granulation.

A number of physicians have discovered that levulose is more readily tolerated by diabetics than any other sugar and

Tupelo has been recommended to many thus affected with wonderful results. It should not be used, however, without the attending physician's investigation and approval.

Another problem of the Tupelo Honey producer is one of early pollen. Many of the keepers find it profitable to move their hives to points in South Georgia, where plenty of natural pollen is available. In fact for months the bees are subjected to an unconscious process of preparation for the brief period of tupelo flow, which in normal seasons is at its height from the middle of April to the middle of May. The flow lasts between three and four weeks according to climatic conditions and the hives are robbed two or three times, practically all of the honey being removed the last time.

Usually during the first part of January, the bees are brought back from their winter quarters in Georgia and they begin to feed almost at once on titi, maple, ironwood and a variety of other early blooming plants. Having been practically dormant for the past three months they are in their weakest condition at this time. During the remainder of January and all of February they are carefully built up and nurtured in preparation for the real work of the spring. In unusually cold seasons it is necessary to feed the bees, but normally they find sufficient sustenance among native growths.

In March the black tupelo gum, oak and other trees begin to bloom and the bees, which are now in good condition, begin to work in earnest. The colonies are encouraged to continue building up and the foundation is placed for the top boxes. At the end of the black tupelo flow and just before the white tupelo blooms, the hives are completely cleaned out, so that the white and dark tupelo may not be mixed. Black tupelo is known to the trade as amber and is sold to manufacturers of candy and confections.

About middle April, the white tupelo flow is at its height and the bees have reached their best condition of the year and they need all their strength, for within three or four weeks many hundred thousand pounds of honey are gathered. The bees work so frantically that the average life during this flow is twenty-one days. They wear out their wings in that time and die.

At the conclusion of the white tupelo flow, some of the producers leave their hives to be filled during June and July with honey and pollen from the wild grape vine and snow vines for the winter months, as all of this is dark honey and not profitable commercially. The most profitably operated apiaries follow a different plan. They screen over and close their hives and transport them into the farming sections of extreme North Florida and South Georgia where they are allowed to pass the rest of the summer in gathering honey and pollen for the winter months. With the arrival of cold weather they become dormant and as stated, in January are brought back to the home apiary to begin the operation all over again.

From the foregoing it is readily ascertainable that the production of tupelo honey does not follow the same smooth roads as that of other varieties. The problems of transportation north and return, the location of the apiaries with reference to owners' homes, as well as the ordinary expenses and replacements incidental thereto, all these make necessary a price slightly higher than for other grades. When one considers the merit of the product, the difference is entirely negligible. The beekeepers who produce Tupelo Honey, during the quarter of a century of its existence, have never striven for riches, but have been, and still are, perfectly satisfied with a fair return for their labor. Their excess profits are in the associations incidental to their work, the beauty and soul's satisfaction of the woods and waters.

It will be remembered that Tupelo Honey is never sold in the comb, but always in liquid form. This gives an essentially purer product as every drop is strained. The honey men have always been proud of their product and taken keen interest in preserving its reputation. To this end a little over a year ago, a Cooperative Association was formed among the most progressive of the beekeepers to perpetuate the progress and purity of Tupelo Honey, as well as to take charge of the marketing of the product.

A GRAY, April morning—cold and dreary even on a palatial extra fare train rushing across the continent. Travel-weary passengers drift into the dining car, scowl at the menu and stare gloomily at the cloud veiled landscape. The waiter deferentially suggests to one, "And will you have honey with your waffle, sir? It is the very finest honey made—pure white tupelo. Yes sir! I'm sure you will like it."

The breakfast is served, and in due time a small, squat jar of crystal clear, pale yellow fluid appears before the weary guest. Its contents are revealed as a delicately flavored, infinitely smooth, slow-pouring liquid, which becomes subtlety itself on the palate, perfect in flavor and consistency. The guest, suddenly hungry, consumes the last drop with satisfaction.

Two thousand miles from the chance diner and his pleasantly, though expensively gratified appetite, there lies a heavily timbered, sparsely settled region of which he never heard, and through it runs a calm, purposeful river with a long Indian name that would be only a jumble of the alphabet to him. It is a friendly river, but it is businesslike and as it rounds a deep curve in the shoreline it neither repulses nor urges one to follow. Yet if one descends the gentle slope of the shore to a boat waiting among graceful, gray tree trunks that stand in the shallow backwaters, there would be no delay in paddling out into the bayou, clear of the clustering trees, past the steamboat landing and out into the current. For those who listen to rivers know that this one has something to say.

Rapidly, happily the miles flow past. Evenly, unhurriedly the river swings on between banks massed with the glorious green of a virgin forest, rich in realization of a southern April. Cypress, cottonwood, water elm, sycamore, laurel oak, cedar, hickory, live oak, chinquapin, water ash, sweet bay, box elder—all these and more crowd its banks and form a background for thickets of willow, button bush, black haw, titi and hackberry. Darkly massed behind them loom giant magnolias dotted with early bloom that trails its exquisite fragrance on the morning air. Wild Wisteria scrambling adventurously over shrubs and trees, swings its first purple tassels in the river breeze, and feathery cottonwood and fluffy willow blooms drift lightly down through the soft air.

Far more numerous than any of these, however, are thickly branched trees with sturdy gray brown trunks and dark, glossy leaves. They seem to be everywhere—tender slips at the water's edge, thick bushy younglings mingled with the forest growth on the low shore, mature trees standing in the still backwaters and lagoons. This is the tupelo gum tree of the southern lowlands. From its branches at this season depend thousands upon thousands of small fuzzy bolls or blooms, on long stems and in thick clusters. And upon those has been founded, casually and gradually, an industry that offers to discriminating world markets a valuable commodity in the form of a choice type of the most wholesome sweet known.

For miles down the river there is no sign of human habitation, but hidden in the edge of the leafy screen along the banks one unwittingly passes many well tenanted homes of tireless, eager workers. Though the air be heavy with the scent of spring blossoms, these busy swarms of Italian bees pay no attention to any but the white tupelo blooms, and the riverfront and swamp in all directions are astir with them through the daylight hours. The "flow" is on: It is tupelo time.

For those who think of Florida only as a tropical winter playground where a fortunate few may loll in summer attire on white sand beaches, there is a revelation in a trip to the little known northwest section of the state. Here four counties dip down to form the last descending point of land before the Gulf Stream sweeps up to hollow out the great curve of the peninsula's western shore. Here is a land underdeveloped drowsing happily among its riches, covetous of no one, desirous of nothing, unselfish to a fault. Endless acres of cutover pine land, worked out years ago by the great lumber companies, are abandoned to pasturage and casual turpentining of the younger growth timber. Deep swamps, thickly crowded with hardwood trees as yet spared the timberman's axe and saw, shelter birds and game in great numbers.

Centrally located in this undeveloped region and fronting on the Gulf of Mexico is Gulf County, created from the southern part of Calhoun County in 1925. It is sparsely settled, there being perhaps no more than 5,000 people in the entire county. Wewahitchka, a small village located in the north central section, is the county seat and is the nucleus of the tupelo honey industry of northwest Florida, with an annual production of 535,000 pounds of fancy white tupelo honey, which brings the producers about $60,000.

Fancy white tupelo honey is considered the choicest kind and grade offered to the trade, as it is delicately flavored, crystal clear, light in color, smooth in consistency, high in density and is not variable in any way. In addition to these advantages the pure white tupelo honey has the remarkable qualities of never granulating and never becoming rancid. One producer at Wewahitchka has a sample of honey which he has kept for nineteen years. It is kept in an ordinary glass jar with a cork, and retains the same flavor, color and consistency which it had in the beginning. Despite these exceptional qualities, white tupelo honey rarely reaches the consumer in an unadulterated state, because the producers for the most part sell direct to canners and commission men who have utilized it to build up and improve blended honey from other sections. The advantage to the concern which bottles honey is obvious; the addition of a small quantity of white tupelo honey to that of other flavors and grades improves the taste and lengthens the time during which it will keep without granulation or deterioration. The disadvantage to the producer who has so carefully handled his apiaries throughout the year in order to guarantee the purity of his tupelo honey is also obvious, since few consumers ever obtain his produce in an unadulterated state or know its source. The remedy, apparently, lies in a movement now on foot to revolutionize the prevailing system of marketing.

The tupelo gum tree, both white and black, is native to the swamps and river bottoms of northwest Florida and grows profusely in them. It also grows in Louisiana, Mississippi and other southern states, but Gulf County apiarists state that the production of pure white tupelo honey has not been reduced to an exact science except in their locality. The black tupelo makes a darker and less desirable honey than the white, and mixing of the two is carefully avoided in the Wewahitchka section where beekeepers have learned to manage their hives in such a way as to accomplish this.

The Chattahoochee River, rising in central Georgia, flows south to the Gulf of Mexico, and is joined near the Florida line by the Flint River from Alabama. From this point the stream is called the Apalachicola until it reaches the Gulf at the town and bay of the same name. For about sixty miles of its lower course the banks and backwaters of the stream are heavily wooded with the tupelo gum, and the river swamps in which this tree thrives vary from one to twenty miles in width. Learning early of the superior quality of honey produced by the tupelo gum and the preference of the bees for it, local apiaries placed their colonies of bees on the river bank or deep in the swamps, often locating ten, twenty or more miles from any human habitation. There are few roads in this section and many apiaries are inaccessible except by boat. Most of the tupelo acreage is leased from its owners by apiarists, though some own the land on which they operate. There are twenty-eight of the larger apiary sites averaging twenty-five acres to the site, and covering more than twenty thousand acres in all. Scientists have stated that bees will fly three miles for honey, but practical apiarists in the Wewahitchka section believe that two miles is an average distance of flight, and they locate their colonies with this in view. The Italian bee predominates in this district, though some of the wild black bees which abound in Florida forests have mingled with hives in a few apiaries. The wild bees are difficult to handle and are not desirable for commercial use.

Honey producers were alarmed and distressed several years ago because of the entrance of cigar box manufacturers into the white tupelo section and the purchase of tupelo gum timber by them. It was found, much to the relief of the apiarists, that the wood of the tupelo gum is too light and brittle for use in box-making and other hardwoods were substituted.

The tupelo gum or cotton gum tree, is usually fifty to seventy-five feet in height and two or three feet in diameter, and it frequents swamps and inundated areas. The base is often enlarged, and the tree has a fairly straight trunk covered with thin, gray-brown bark, deeply furrowed. The branches are smooth and light brown, and the slender, pointed leaves are thick, their upper surface being dark green and lustrous and the lower pale and downy. The blossoms are usually borne on separate trees, the male in dense round clusters and the female alone on long slender stems. The bloom appears before the leaves on the black tupelo gum, but the opposite is true of the white tupelo. The male tupelo bloom resembles a black clove and is said to contain more honey than the female bloom, which is a small fuzzy ball. Each of them secretes nectar constantly and profusely from twenty to twenty-five days, and bees return again and again to the same blossoms for honey, which often gathers so thickly that it could be scraped off with a knife. It is believed that twelve days elapse from the bud to the full bloom of the tupelo, and after the period of secretion the pod turns brown and drops off.

The present State apiary inspection for that district has resided near Wewahitchka since 1885, and he has records of carefully conducted tests in which single colonies of bees have been known to gather twenty pounds of honey in one day. In a favorable season one apiary containing ninety colonies produced thirty-eight barrels of honey in three weeks, each barrel containing thirty gallons. The average production of one hundred colonies during the brief period in which they gather white tupelo honey is twenty, thirty gallon barrels, but records of twenty-five and even twenty-seven barrels are common. The confinement of the bees' activities to the short space of three or four weeks makes possible the production of unadulterated white tupelo honey, and the insects "on vacation" during the remainder of the year.

The largest individual producer in this section has an apiary thirteen miles from Wewahitchka where 326 colonies of bees average 40,000 pounds of pure white tupelo honey each season. The presence of high water in the tupelo swamps during several months of each year renders it necessary to build many of the apiaries on platforms fourteen to sixteen feet in height and three hundred to seven hundred feet long. The honey house containing two stories, is built immediately behind the platform at its center and an inclined runway leads from each story to a small wharf or steamboat landing.

The hives are placed in double rows along the platform, with a passageway between, and the entire work of harvesting the honey and packing it for shipment is handled in the honey house at each apiary. All white tupelo honey is sold in the extracted form. When the hives are robbed the combs are brought into the upper story of the honey house and placed in a large vat, where a slicer removes the caps. It is then placed in frames in a revolving drum and the honey is extracted by centrifugal force, after which it runs through a pipe into a large tank of very tight construction on the lower floor of the honey house. Here the small amount of sediment and foreign matter contained in the honey settles and the finished product is drawn off into barrels constructed for this particular purpose. Because of its weight, honey is particularly subject to leakage, and it is difficult to handle in bulk. The barrels used are specially coopered of choice cypress, carefully washed, dried and pariffined inside. They are used only once, each season's shipments going out in new barrels. River steamboats run twice a week and the barreled honey is delivered to them direct from the dock at the front of each apiary, or from regular landings.

Prices received by the producers are very low in comparison with the high price finally paid by the few consumers who obtain this choice product in its unadulterated form.

For many years the honey was bid in by representatives of large commission houses who come to Wewahitchka for that purpose at the close of the honey harvesting season each spring. Eventually the monopoly which a few of these held forced the price so low that local producers refused to sell and formed a cooperative association which has successfully handled the crop in recent seasons.

The advantages and possibilities of the industry are obvious. The apiaries require comparatively little attention, though practical operators are constantly studying the needs of the industry. The net returns on each producer's investment are good, even at present low market prices. It is, however, a seasonal business, involving very heavy work during the harvesting season and slack periods of employment at other times. Because of the isolated location of the apiaries, losses from forest fires and similar sources are considerable. Ill-timed rains sometimes prove very costly to honey producers, and a single hard shower in the height of the white tupelo flow is estimated to cost the producers $25,000 or more. Apiaries have been carefully spaced with regard to the probable number of bees operating on each tract, and as yet the white tupelo is plentiful and there has been no shortage of honey material.

The industry was established in this section more than fifty years ago, apparently in an accidental manner, and it has grown to proportions which are admittedly beyond the capacity of local producers to handle satisfactorily. It remains only for business to recognize the possibilities of the industry and exploit them through practical channels, in order that the public at large may come to know by name a delicious American product now enjoyed by only a few—fancy white tupelo honey.

Beekeeping in Florida

By

J. J. WILDER

Foreword

The many requests for information received by the State Department of Agriculture have shown that a large number of people are interested in the possibilities of beekeeping in Florida. Requests have been received not only from people residing in Florida, but also from people living in many other states. A number of the people are interested in beekeeping in Florida merely as a pastime—an activity at which they can enjoy their spare moments. Others, however, are interested in beekeeping because of financial returns, either as a sideline or on a large commercial scale.

It is for such beginners in beekeeping that this bulletin is written. The author, Mr. Wilder, has had many years' experience with bees in practically all sections of Florida, and at the present time he owns about 10,000 colonies of bees. He is, therefore, unusually well qualified to inform the prospective beekeeper in Florida as to the best procedure in beginning his apiary.

NATHAN MAYO
Commissioner of Agriculture

Beekeeping in Florida

One of the first apiaries of any consequence in the State was established on the Florida East Coast on the west side of the Halifax River, where the city of Daytona now stands. This apiary was established in 1872 by a New York company which was in that section producing lemons and oranges. The production of lemons, oranges and honey made a very good combination. The company would come southward during early fall in time to gather their fruit and honey. After spending a few months in Florida, they would sail back to New York City in the spring with a cargo of Florida fruit and honey. This practice excited considerable attention around New York as well as in certain Florida towns.

Probably the next apiary of any importance was started near the city of Wewahitchka in Gulf county by Mr. S. S. Alderman, who also grew oranges along with the production of honey. Just a little later Mr. W. S. Hart, located at Hawks Park in Volusia county, began producing honey and fruit in like manner.

This early development of beekeeping in Florida took place between 1872 and 1888. There was not much to Florida at that time. The pioneer beekeepers had a hard time of it. They obtained their bees from the forest, lived in remote sections of the country which could be reached only by small vessels, and were seldom visited by those from other parts of the country.

The success of S. S. Alderman and W. S. Hart soon caused reports to be widely circulated that an average of one barrel, or four hundred pounds, of honey per colony was being secured in Florida. This report meant much to Florida in beekeeping for almost at once people began to establish apiaries all over the State and to put in modern equipment. Progress has continued down to the present time.

THE BEE

It is generally known among beekeepers in the southeast that Florida has a black bee which has thrived in the forests of the State for many years. These bees still exist in the State and can be found in the large cypress timber of the Everglades, the Okeefenokee swamp, and the heavy timbered sections in the western part of the State. Just when or who brought the first bees to Florida is not known. On this subject, Mr. Jas. I. Hambleton, apiculturist of the United States Department of Agriculture, writes: "The most authentic record states that the black or German bees were introduced into West Florida not later than 1763. In all probability the honey bee occurred in East Florida before that, as black bees were introduced in New England as early as 1638. William Bartram, describing a journey taken in 1773, says that honey bees were numerous all anong the Eastern Continent from Nova Scotia to East Florida. He further states that honey bees were common enough in forests so as to be thought by the inhabitants to be natives of this continent."

The movements of this wild bee in Florida are quiet, and no bee is as busy on flowers as it is. The activity of these bees is far beyond the common bees, and they are very cross and quick as lightning to sting. When a tree containing these bees is cut, they act about like hornets disturbed from their nest. They produce a large amount of honey per colony, yet they do not seem to adhere at all to the idea of being domesticated. They are not contented to live in hives and will desert them time and time again for the forest. Only in a small measure do they adhere to our modern methods of handling bees. The bees are also so furious that they are not desirable to have around a farm. The very presence of a human being seems to completely demoralize them. In many cases the comb they build has irregular cells, yet they cap their honey beautifully white, and it is of good flavor like that produced by other bees.

The Italian and Caucasian are the more domesticated bees, and these two races predominate in the commercial apiaries of Florida. The Italian is particularly desirable for the production of extracted honey, while the Caucasian excells in the production of comb in shallow frames or sections. Many small beekeepers in the State still keep the black or German bee, but the two races just mentioned are much more prolific and desirable for the many different honey flows.

WHERE TO KEEP BEES

This question can readily be answered. "Bees may be kept in Florida anywhere you live, or are moving to." There are no barren spots in Florida so far as beekeeping and honey production are concerned. This does not mean that all sections of Florida afford good bee pasture at all seasons of the year. It

Showing Part of a Large Apiary in Palmetto, Near Tampa, Florida

does mean that there is no large area in the State but what at some time during the year will furnish bee pasture. One must be careful, though, to see that the hives are placed in some thinly shaded place where they can be properly watched and taken care of. Should one be going into beekeeping on a commercial scale, it is necessary, of course, to consider transportation, the kind of honey plants that are available, etc.

BEGINNING THE APIARY

The right start in beekeeping means much toward success. At the very beginning the apiary site should be selected, and this done with great care and consideration.

Bees should never be kept near stock where there would be danger of horses, cattle, hogs, etc. being stung by them. As a rule, all animals understand to stay away from bees, and they will usually do this if they have their freedom. The apiary should be far enough away so that there will be no danger of either man or animals being stung, yet it should be near enough to the house so that it can be closely watched. It is advisable for someone to visit the bees rather often, for bees will soon become familiar with people who pass by. After the bees become familiar with people, there is little danger of a volunteer attack of the bees or any stings from just passing among the hives.

The location should be thinly shaded, but never should there be a dense shade overhead. A dense shade will cause the hives to be more or less damp, especially during rainy weather, and this is detrimental to the bees. The dampness also causes the hives to decay more rapidly. No shade at all would be preferable to a dense one.

The first colony of bees should be placed in the site selected. As fast as an increase in made, the hives should be lined up about four feet apart so as to give sufficient room to work around each. The rows of hives should be at least ten feet apart so that if necessary a truck may pass between the rows. It is best to let the hives face southward, although southwest or southeast will do. It is necessary to place the hives on stands some twelve or eighteen inches high so that the ground about them can be kept free of litter and vegetation.

As soon as there are a few hives in the apiary, a suitable,

Showing Part of a Large Apiary in Orange Grove

neat, small honey house or room should be erected close by the side of the apiary. It is preferable to locate the honey house on the side of the apiary nearest the residence so that it may be visited without passing among the bees. The honey house may serve as a workshop as well as a packing and extracting room when the honey crop is ready. Honey is to be kept in this room and only enough carried to the residence for a meal or so at a time. Honey tends to toll in bees and other insects and often makes a rather messy job to keep clean. The honey house is the place for it and it can be readily removed when needed for the market or table. An extractor, uncapping tank, storing tank, and a large work table on which to pack the honey are needed in the honey house.

OBTAINING BEES

There are bees in every nook and corner of Florida, and one should have no trouble in obtaining a start almost at his very door. It is not necessary to send north or west for bees, as they can be obtained in Florida. Bees in Florida are inspected as to disease by authorized State inspectors, and they will see to it that the bees are free from disease. When bees are secured from outside the State, it is impossible to know just what one is obtaining, and it may later be discovered that the bees are diseased.

As already stated, it is advisable to obtain pure Italian or Caucasian stock, and possibly better than either is the Caucasian-Italian stock crossed. The bees purchased should be in either eight- or ten-frame modern standard size hives. It one expects to produce extracted honey, the ten-frame hives and pure Italian bees are recommended. If one expects to produce chunk honey or comb honey in one-pound sections, bees in eight-frame hives should be secured. It is preferable to get either Caucasian-Italian or Caucasion stock for producing chunk or comb honey, as these two varieties are about the best comb builders and they cap their honey beautifully white.

For each hive, three regular shallow extracting supers should be purchased if one is going to produce either extracted or chunk honey. If comb honey in sections is to be produced, then two supers are all that one needs. The best equipment obtainable with full sheets of foundation in all frames and sections should be used by all means. One must see that all hives and hive parts are properly set up according to instructions given in the bee supply catalog. If this is not done, it will be found out later, much to one's sorrow.

HONEY YIELDS

The yield of honey per colony will vary for different sections of the State. The variations will depend almost entirely upon the supply of honey plants in each section. The State as a whole will probably average 50 to 70 pounds of extracted honey, although there are a number of localities that will average up to 100 pounds of extracted honey per colony. A few exceptional areas may be found where the average is as high as 200 pounds of extracted honey per colony.

To express it in another way, it may be said that in the Tupelo Gum region of West Florida, the average per colony is about 100 pounds of extracted honey; in the partridge pea region, about 60 pounds per colony; and in the saw palmetto region, about 50 pounds per colony. The sunflower region as a rule gives the best yields, sometimes averaging as much as 200 pounds of extracted honey per colony. Then in the Black Mangrove region the average is often around 150 pounds per colony, while in the gallberry region the average may be as low as 40 pounds of extracted honey per colony.

GRADING AND PACKING HONEY

Whether an apiary has one colony or fifty colonies, the beekeeper should know how to properly grade and pack honey even for his own table, and especially all he expects to put on the market. The surplus honey should never be put up in just any kind of container, but it must be correctly put up in good honey containers.

Honey produced in Florida, as a rule, has a good flavor and good color. Sometimes, however, it is a little thin in body even after it has been left in the care of the bees until it is well capped over. The bees cap the honey when it is finished, but as a rule they do not do this until they have given it the body they intend it to have. One should remember that honey, when first gathered, is nothing but sweet sap of the honey plants, thin, devoid of flavor, and quick to ferment until well evaporated.

At the present time the demand is greatest for honey put up in retail containers. The one-pound square jars have been found most suitable for the best grades of both chunk and extracted honey. The two and one-half-pound cans are best for grades just a little off in color. The next size is the regular five-pound honey pail. Syrup pails will not do as they are too thin and frail, and the friction top does not drive in sufficiently tight to remain and not leak. The off-grade extracted or comb honey can be put up in regular honey pails or in two and one-half-pound glass jars.

Extracted honey should be well strained before it goes into the storage tank. It should be allowed to remain there for several days so that gravitation will clear all matter from the honey, then it can be drawn off into the containers and sealed up at once. All packages can be neatly labeled under your own signature, together with the guarantee and net weight.

It is generally advisable to put up some of the honey with comb and some without comb. One can often sell ten times as much packed comb and extracted honey together as straight extracted honey alone. Many people want comb in their honey in spite of whatever they may think best. In packing comb from the regular shallow frames along with extracted honey, one must be careful to put in as large pieces as possible and never chip up or put up little trimmings. It is desirable to let the honey appear in as large pieces as possible. The pieces should be suspended so that they will stand up; they should not be put in flat, for honey naturally looks better from an end view than from a side view. One must remember to cut out only tender young white comb and to place the fancy crop in glass containers.

BLENDING HONEY

It is a well known fact that practically all the extracted honey on the market is blended (not compounded) from several sources. Blending is done for several reasons. First it makes a better table article because the flavor of blended honey is a combination of the flavors of several different kinds of honey. As most people are aware, the flavor of honey is governed by the plant from which it is made, so that blended honey combines the different flavors. All real honey lovers will agree on this point. The honey may be blended just as it comes from the extractor, or on the table when cutting the comb.

Blending honey has reference only to the very best honey and not to any of inferior quality. A poor grade should never be in a blend, or it will ruin all. It is better to put the cheap honey up separately and sell as such. This applies to both the color and flavor of honey. Some poor honey has a fine color, and some very fine honey has poor color. It is seldom if ever advisable to blend dark honey with light, or honey of poor flavor with that of good flavor, but a blend should always be with honey of similar color and quality of flavor.

The blending of honey is particularly important in Florida because there are a great many kinds of honey coming along during the season. Often one honey flow comes in very close behind another flow, and this happens so frequently that there is very little honey produced in Florida which is pure as to source. It is all blended more or less by the bees themselves, for sometimes a single comb will contain three or four different kinds of honey.

Blending honey not only makes it a better table article, but the greatest advantage is that it stays granulation. Much of the Florida honey, especially that produced in the southern part of the State, will granulate. The honey in the western part of the State, particularly in the great White Tupelo Gum region, does not granulate easily. If a large percent of non-granulating honey is blended with the honey that granulates, then granulation is stayed, often indefinitely even on the northern markets. There is enough non-granulating honey produced in Florida, if properly blended with the granulating honey, to keep all in a liquid condition.

Florida therefore has the opportunity to put up honey in its natural state that will keep without granulating, which eliminates the necessity of heating the honey to make it keep. Honey that is sold with the guarantee that it will not granulate is more in demand, for no honey buyer outside of a bottler wants table honey to turn to sugar or candy.

WINTERING BEES

To those less informed the winter care of bees in an almost tropical country like Florida seems of little importance, and perhaps is far less important than in other parts of the country. Some special care, however, is needed by the bees during the winter months even in Florida.

During the first part of the winter, the bees should be looked over carefully and even the queens and their work of egg-laying noted. Some honey is generally coming in at this time, as the weather is usually still warm enough to allow the bees to work. The first part of December is the most opportune time to make the examination because old and failing queens may be easily detected at such time by the strength of the colonies and size of the brood nest. A good queen at the beginning of the winter season should be laying well with plenty of young bees in the colony; if this is not the case, then the bees should be re-queened.

While the cover is off and the queen's progress being noted, it is advisable to see about the stores in the super just above the brood nest. This super should be full or nearly so of sealed stores. The bees may not draw very heavily on the honey the first part of the winter, but the latter part they will because they are rearing so many young. The cover to the hive should be a good one that does not leak, and the bottom board must be sound. It is also important to see that the hive is on a good foundation.

The colony with a good queen and plenty of stores is ready for the winter and will need no further care or attention until spring. Plenty of stores above a good queen is highly important; otherwise, losses from starvation are almost certain, or the colony will be too weakened from lack of honey to keep up the raising of young bees. One must not forget that bees will perish during cold weather even in Florida where winters are short and generally mild, unless they are given sufficient care.

PERPETUAL HONEY FLOWS

The question is often asked, "Can I keep bees in Florida and have a honey flow the year around?" The idea is to have a honey flow twelve months in the year, taking honey off, packing, raising bees and queens, etc., the year round. As a general rule, however, nowhere in Florida can one depend upon such a condition year after year. All of Florida is subject to cold snaps, light frost, and once in a while freezes, which to a large extent play havoc with vegetation. This would mean disappointment to the beekeeper who is expecting to run his honey extractor or pack honey every month in the year. Some years this can be done, but years when light frost and freezes come around this cannot be done.

From coast to coast across the peninsula for about one hundred miles, taking in the section where Lake Okeechobee lies, there are large areas of pennyroyal, a winter-blooming nectar plant that gives a good and reliable flow of nectar from the time the goldenrod ceases to bloom on through the winter months until citrus begins to bloom. This is ideal for honey production, bee and queen raising, but even here this is interfered with by sharp cold snaps.

Through the section just mentioned, the average per colony is far greater than elsewhere in the State. This is simply because there are more honey plants and a nearer perpetual honey flow with only a few days intermission from one to another. This section embraces, of course, a large area in the extreme southern part of the State. Honey extractors can be seen running in various places through this section during November, December, January and February. Often the number of bees will increase during these months, and queen bees reared and mated.

This is perhaps the most favored area in Florida for beekeeping in all its branches. Pennyroyal is the greatest yielder during these months, yet there are other honey plants that come along and bloom during the same period which add greatly to the flow of honey and abundance of pollen.

REMOVING THE HONEY

The surplus honey of any colony can be removed at any time, but beyond this no honey should be taken. Because one sees blooming flowers almost twelve months in the year around over Florida is no reason why they are real honey plants and the bees can gather honey from them. Therefore, sufficient honey should always be left for the bees to live on.

It is important to keep a close watch on the bees so that they will not have any more storing room than needed. The bee moth will actually eat up the comb in a normal colony of bees if there is so much storing room that the bees cannot properly care for the hive by crawling over it and removing the eggs or tiny larva of the miller that lays the eggs. It is a common sight, and not a good one, to see a hive of bees with the combs all destroyed in the top by the moth. Bees should have only the proper amount of room at all times, but most particularly at times when there is no honey flow and breeding may be at a low ebb. A close watch must be kept on the bee moth or it is apt to cause great loss of comb.

When a honey flow starts, it is necessary to look out for super room and keep just enough storing room ahead of the bees so that they can fill up all supers by the end of the honey flow. Too much would be detrimental and not enough would be a loss. To this end every colony should be closely watched and visited every week to see that all are kept supplied with storing room. When the honey flow goes off, then all the surplus honey can be removed, packed, and placed on the market. One super, however, must be left full or nearly full of stores for the use of the bees.

FRAME MANIPULATIONS

All modern hives have loose hanging frames in which the bees build the comb, live and rear their young. Every colony should be examined carefully each week, or at least every few weeks. Each comb in the bottom story of a hive should be examined to see whether there are enough brood eggs of the queen and a sufficient amount of honey.

If there is no honey in the super, it is necessary to supply a frame of honey from some heavy hive. If there are not as many bees in some colonies as in others, one may take a frame of capped brood from one of the strongest and best colonies and place it in the weaker colony. In this way the weaker colonies can be built up. If no brood is seen or the colony is growing very weak, the hive may have a poor queen or none at all. Such colonies should occasionally be given a frame of brood in all stages of development, which will enable them to grow stronger and raise a queen from the brood given them. Or, in the meantime, one may order a queen and introduce her into the colony, which may often save a colony from a downward drift or perhaps a total loss.

Frame manipulation is of the greatest importance in beekeeping, for right here the wheel of fortune in beekeeping may turn. This is particularly outstanding in changing combs as just mentioned above.

A Friendly Beeman

SWARMING

It is not customary even among beginners and small beekeepers to allow the bees to swarm naturally, as much better results are obtained when the swarming is controlled by the beekeeper. The operations of increasing colonies and controlling swarming are both done with one stroke. When a very strong, heavily populated colony of bees is properly swarmed once each season, that colony and the one made from it are both cured of the swarming fever for the year.

The strongest colonies should be divided up into equal parts, in the early part of the year, some three or four weeks before natural swarming time. This means taking from the old hive one-half the bees, one-half the brood, one-half the comb, and one-half the honey. As the hive is being divided, one should look for the queen. The frame of comb on which the queen is found should be put with the half that is to make a new hive. The bees in the old hive can raise themselves a new queen, although it is often preferable to buy a queen for the queenless half.

The operation is not a success unless the queen is put with the new stand, because if the bees that are carried away to a new stand find themselves queenless, they will boil out of the hive, pry about looking for the queen, and invariably go back to the old stand in an effort to find their mother. This depopulates the newly made hive, but if the queen is there the bees will not leave her. The old half of the hive will have no idea where their mother has departed to and will at once set out to raise another, or will readily accept a new queen if one is introduced.

This is simple and easy when everything is in readiness, and it can best be done late in the afternoon by those inexperienced in the operation. The bees will thus be given over night to satisfy and content themselves, while if done in early morning there will be turmoil all day among the two divisions, the bees on the old stand looking for their mother and the bees on the new stand making their new home. Before this is done a new empty hive for each colony must be properly prepared, and the frames should contain full sheets of foundation or ready built comb.

When the division is made, there should be four or five frames in each hive of ready built comb containing brood and honey. This is supplied when the division is made, but a space should be left without any comb on one side of each of the hives. The frames containing full sheets of foundation from the newly prepared hive should be inserted in these spaces. One frame containing foundation can be placed right in the middle of the ready built combs in each of the divisions. This will give the bees some comb to build and they will start at once to draw out the foundation. As fast as they draw it out, the queen will fill it with brood and one will soon have solid slabs of brood.

On every visit, a frame of foundation should be inserted in like manner until a full set of combs are drawn out, then all one has to do is to keep the bees properly supered and two hives rather than one will be making the honey.

Increases during any time of the year can be made in like manner, but only with strong, heavy colonies. The weak colonies and those of medium strength naturally have a struggle to exist and to divide them would mean disaster and great loss.

To become successful in beekeeping, one must study the nature and habits of the honey bee in order to learn the best methods of bee culture. An effort should be made to learn about the plants upon which the bees feed. A number of good books are available on bee culture, which can be obtained at a reasonable price. There are also a few monthly publications on bees that contain valuable information. Whenever possible, the prospective beekeeper should visit one or more progressive beekeepers in the locality in which he intends locating and watch the methods of handling bees. The more information one can secure, and the better it is applied, the greater will be the chances for success.

LAWS RELATING TO HONEY BEES
Florida Statutes Chapter 584

584.01 Powers of state plant board over honey bees.—

The state plant board of Florida may deal with American and European foul brood, Isle of Wight disease and all other contagious or infectious diseases of honey bees which, in its opinion, may be prevented, controlled or eradicated; and may make, promulgate and enforce such rules, ordinances and regulations and do and perform such acts, through its agents or otherwise, as in its judgment may be necessary to control, eradicate or prevent the introduction, spread or dissemination of any and all contagious diseases of honey bees, as far as may be possible, and all such rules, ordinances and regulations of said plant board shall have the force and effect of law.

584.02 Certificate of inspection to accompany shipments.—

All honeybees (except bees in combless packages) and used beekeeping equipment shipped or moved into the State of Florida, or shipped or moved within the State of Florida, shall be accompanied by a permit issued by the plant commission of the state plant board of Florida. Before any bees (except bees in combless packages) or used beekeeping equipment is shipped or moved from any other state into the State of Florida, the owner thereof shall make application on forms provided by the plant commissioner of the state plant board of Florida for a permit. The application shall be accompanied by a certificate of inspection signed by the state entomologist, state apiary inspector, or corresponding official of the state from which such bees or equipment are shipped or moved. Such certificate shall certify that all of the colonies, apiaries, and beeyards owned or operated by the applicant, his agents or representatives, have been inspected annually at a time when the bees are actively rearing brood, including one inspection within the period of thirty days immediately preceding the date of shipment or movement into Florida, and that no American foulbrood or other contagious or infectious diseases have been found in any colony, apiary, beeyard or other places where bees or equipment have been held by the applicant, within the period of two years immediately preceding the date of shipment or movement into Florida; provided that when honeybees are to be shipped into this state from other states or countries wherein no official apiary inspector or state entomologist is available, the state plant board of Florida may issue permit for such shipment upon presentation of suitable evidence showing such bees to be free from disease.

584.03 Authority of plant board to enter depots, etc., to make inspections.—

The state plant board, its agents and employees, may enter any depot, express office, store-room, warehouse or premises for the purpose of inspecting any honey bees or beekeeping fixtures or appliances therein or thought to be therein, for the purpose of ascertaining whether said bees or fixtures are infected with any contagious or infectious disease, or which they may have reason to believe have been, or are being transported in violation of any of the provisions of this chapter.

584.04 Plant board may require removal, destruction, etc., of exposed or infected bees.—

The plant board, through its agents or employees, may require the removal from this state of any honey bees or beekeeping fixtures which have been brought into the state in violation of the provisions of this chapter, or if finding any honey bees or fixtures infected with any contagious or infectious disease, or if finding that such bees or fixtures have been exposed to danger of infection by such a disease, may require the destruction, treatment or disinfection of such infected or exposed bees, hives, fixtures or appliances.

584.06 Penalty for violation.—

Whoever violates any of the provisions of Chapter 584, Florida Statutes, or whoever violates any of the rules and regulations promulgated by the state plant board in accordance with the provisions of said chapter, shall, for the first offense be deemed guilty of a misdemeanor and upon conviction thereof be punished by a fine of not less than one hundred dollars nor more than five hundred dollars or by imprisonment for not more than six months in the county jail, and upon a second conviction thereof shall be deemed guilty of a felony and shall be punished by imprisonment in the state prison for a term not to exceed three years. It shall be the duty of the sheriffs and the Florida highway patrol officers to enforce the provisions of this chapter relating to the movement of bees and used bee equipment into the state as well as movement thereof within the state.

NOTES ON THE POLLINATION OF SOME SUBTROPICAL FRUIT PLANTS

D. O. WOLFENBARGER and MILTON COBIN

University of Florida Subtropical Experiment Station, Homestead

Introduction

Cross pollination is a requisite for some of the subtropical fruits while in others it is of no consequence. Insects are necessary for pollen transfer in some cases, while wind is the agent of pollen transfer for other fruits, or complements insect pollination in other plants. Other factors sometimes adversely affect the set of fruit. These factors include the following: diseases, insect pests, high humidity and rainfall during the flowering period, low temperatures, and inadequate nutrition. Much remains, however, to be learned about pollination of many economic and potentially economic subtropical fruits.

A very brief summary is given to cover the present knowledge of pollination of our leading subtropical plants with notes on the places where honeybees may occupy useful places in pollen dispersion. In doing this the writings of botanists and horticulturists reporting on different plants have been used freely. The lack of knowledge pertaining to pollination of many plants is great; hence, the need for and justification of the recently initiated research by the Florida Agricultural Experiment Stations.

Annonaceous Fruits

These fruits now occupy a very minor position in Florida horticulture, and are almost restricted to the sugar apple, soursop and the atemoya. The blossoms were regarded as having primitive structures by Eastwood (1943), with lack of attractive color which excluded pollination by honeybees. Moths and flies were considered to do most of the pollinating.

Avocado

Although the avocado has perfect flowers, each capable of producing pollen and of developing into a fruit the pollen-shedding and pollen-reception of each blossom occurs at different periods of time. The avocado exhibits this phenomenon to a greater degree than any other fruit plant known. It is briefly summarized by the statement that the flower is receptive to pollen during its first opening, and that it sheds pollen during the second opening. A period of 12 to 36 hours separates the pollen reception from the pollen shedding.

Insect pollination is considered helpful in avocado production. Honeybees, various flies and wasps have been observed working avocado blooms. Evaluation of the different insects as pollinating agents remains, however, to be determined. It is considered probable that colonies of honeybees in an avocado grove would assist in increasing a set of fruit.

The effects of distance from reciprocating avocado varieties on fruit production have been determined, whatever the effects of winds, honeybees or other factors may have been. This was done by Wolfe, et al, (1949), by counting the fruit produced per tree at different distances from trees of a different pollen source. Curves were drawn from data given, and are shown, Figure 1, for the Taylor and Wagner varieties. The average set of fruit decreased as the distance increased from the pollen source.

There is a principle involved in the above paragraph with reference to the distance factor which is considered important. This principle visualizes a simple measurement of the incidence and distance of pollination under field conditions. This principle was involved in the work reported to this Association by Parris and Haynie (1950) relative to watermelon production. More widespread application of this principle is recommended.

Barbados or West Indian Cherry

The present production of this fruit is confined to scattered specimens or occasionally small groups of specimens located in grove, estate, or back-yard plantings. Honeybees forage extensively on the Barbados cherry and are the principle agents of dispersion of pollen of this fruit according to Cobin (1948).

Guava

Guava plants are very prevalent throughout southern Florida. Rather extensive plantings have been made and the plant grows wild. The development of improved varieties is likely to result in additional plantings. The flowers of this shrubby tree, and many of its family relatives, attract many different insects, among which is the honeybee. Cross-pollination is considered the rule for the guava by Hayes (1945). Several colonies of bees were once located near the center of a 40-acre grove. This was a commercial venture which had for its objective an increased set of fruit. It was planned to determine the fruit set at different distances from these colonies in a manner similar to that reported above for avocados. The grove was abandoned, and the bees were removed so the plan was abandoned. No study is known to have been made on the effectiveness of the honeybee for guava pollination.

Limes

Limes, a species of *Citrus,* according to Frost (1943), are considered as oranges and grapefruit in having no need for pollination in order to increase the set of fruit. Where they are abundant, however, honeybees may obtain nectar from limes, as from orange and grapefruit bloom.

Litchi

The litchi plant produces small inconspicuous flowers, without petals, which are staminate, pistillate, or hermaphroditic. Fruit set is believed to result from pollination, although almost nothing is known in regard to this factor. "Insects are probably responsible for pollination," according to Hayes (1945). Since the acreages planted have been increasing in Florida and since there are now several large scale plantings extensive observations on pollination of the litchi may soon be possible.

Macadamia Nut

The present plantings of the macadamia nut are limited to specimens on estates or in back yards. The future potential of this nut tree is unknown. The macadamia nut is reported by Hartung (1940) to be cross-pollinated, ". . . by insects, principally bees, which visit the flowers at the time when nectar is produced in large quantities." He also believed that pollinators affect the amount of fruit set.

Mango

The mango is considered entomophilous, i.e., pollinated by insects, according to Popenoe (1927) although some writers have reported that the mango is wind-pollinated. Trees which were caged to exclude pollinating insects remained fruitless. Although many flies (blue-bottle and related species, are observed on mango blossoms no recent organized study is known to have been made on insects working on mango blossoms. Honeybees have been observed in small numbers on mango blossoms. When some other plants are near and blossom concurrently with mango trees, honeybees have been observed to seek the pollen of the other plants and to ignore the mango blossoms. This phenomenon has been observed in Dade County in mixed plantings of avocados and mangos. The discovery of some practical means for increasing the amount and regularity of mango fruiting is greatly needed. The problem of fruitfulness of the mango, however, is far more complex than simply a matter of adequate insect pollination.

Papaya

The papaya plant is normally either staminate, pistillate or hermaphroditic, or it may be of bisexual type, or it may be one of various intergrading forms. Pollination is accomplished by insects and wind which disperse the pollen to the pistillate flowers according to Storey (1941). The writers have not observed visitations of papaya blooms by honeybees.

Pineapple

Pineapple planting stock comes from vegetative sources such as slips, suckers, ratoons or divisions of the plant. Short inflorescence stalks grow from these and produce a head of small flowers. Flies are occasionally seen on the flowers. Nothing is recognized of the value of any insect, rain wind or other agent dispersing pollen on pineapple fruiting. Seed development in pineapples is rare.

Literature Cited

Cobin, Milton. 1948. The Barbados or West Indian Cherry in Florida. U. of Fla. Sub-Tropical Experiment Station. Mimeo. Report 14: 1-3.

Eastwood, H. W. 1943. The Custard Apple. Div. of Hort. Dept. Agric. New So. Wales. Unnumbered leaflet: 1-7.

Frost, Howard Brett. 1943. Seed reproduction. Development of gamets and embryos. Ch. 8 pp. 767-815. *In* The Citrus Industry. Vol. 1. Edited by Herbert John Weber and Leon Dexter Batchelor. Univ. Calif. Press. Berkeley and Los Angeles.

Hartung, Margaret E. 1941. Macedonia. Pollination. Rept. Hawaii Agric. Expt. Sta. for 1940: 51-52.

Hayes, W. B. 1945. The litchi. *In* Fruit Growing in India. Chap. XVIII: 180-186. Allahabad Law Journal Press, Allahabad, India.

Paris, G. K. and John D. Haynie. 1950. The effect of honey bees in watermelon fields on set of melons: A preliminary report. The Honey Bees Service to Agriculture. State of Fla. Dept. Agric. Bul. No. 135: 45-49, illus.

Popenoe, Wilson. 1927. *In* Manual of tropical and subtropical fruits. p. 119. The MacMillan Company. New York.

Storey, W. B. 1941. Papaya production in the Hawaiian Islands. Hawaii Agric. Expt. Sta. Bul. 87 (pt. I): 5-22, illus.

Wolfe, H. S., L. R. Toy, A. L. Stahl. 1949. Avocado production in Florida. Fla. Agr. Ext. Bul. 141: 1-124, illus. Revised G. D. Ruehle.

HONEYBEES IN FLORIDA'S PASTURE DEVELOPMENT

By G. B. KILLINGER

Agronomist, Agricultural Experiment Stations

and

JOHN D. HAYNIE

Apiculturist, Agricultural Extension Service, Univ. of Florida

Red clover, White Dutch, Ladino, Crimson, Sweet clover and others depend almost wholly on insects for pollination and seed set.

The flower structure of clovers and other legumes makes wind pollination of minor importance.

The pollination job for an acre of legume blooms is too large for most people to realize. An acre of Red clover contains over 400,000,000 blooms; an acre of White Dutch clover will contain about 800,000,000. Crimson clover will contain about three-fourths as many blooms per acre as Red clover.

The honeybee is the most efficient and best adapted insect for the pollination of crops under Man's control. The uncontrolled natural insects which play a meager role in legume polination are wasps, bumblebees, flies, butterflies, and moths.

Findings in a three-year study by the United States Department of Agriculture in Henry County, Ohio, show Red clover as pollinated 83 percent by honeybees, 15 percent by bumblebees, and 3 percent by all other insects. The general impression has been that Red clover could not be pollinated by honeybees. Data on Red clover pollination in Ohio and Florida disprove this statement.

In Ohio, the average seed yield of Red clover is about one bushel per acre. With the maximum number of honeybees and other pollinating insects, this yield can be raised to 12 bushels per acre. Ohio Experiment Station shows that seed yields are directly correlated with the number of honeybees.

In Florida, farmers are interested in increasing legume seed production and reseeding of legumes in pastures. The number of honeybees and its constancy over a period of years should be of much concern to farmers. For each dollar that the beekeeper receives, $15 to $25 worth of pollination services are returned to agriculture. As long as honey production is profitable to the beekeeper, the farmer will receive free pollination. When the number of bees is increased, seed yields increase, and honey production drops off to nothing. The beekeeper must receive pay in the form of rental instead of a honey crop.

In some areas a good pasture program is carried on where the cattle are removed from the pastures and allow the clovers to bloom and set seed. If there is a large number of honeybees and other wild pollinating insects, the blooms will be visited and seed set. If there is a shortage of natural wild pollinating insects, the farmer is encouraged to own whatever number of colonies he can profitable operate. Fewer colonies are needed to insure pollination in reseeding pastures than when clovers are grown for seed production.

In 1949, the Experiment Stations and Extension Service cooperated in some preliminary studies on pollination of Red clover at Gainesville.

Results for 1949 indicated that very little clover seed would be set by Red clover in the absence of bees. Honeybees increased the Red clover seed production from 11 to 58 pounds per acre.

Florida cattlemen realize that maximum success in their pasture program depend on legumes. Legumes, particularly winter legumes, fix large quantities of nitrogen out of the atmosphere and make it available in the soil for grass growth. In addition to nitrogen fixation, legumes are markedly higher in protein and minerals than grasses, resulting in higher animal production and better finish. With a group of plants that not only stimulate grass production, but also supply a high quality feed themselves at a time of year when the feed supply is critical, it becomes imperative that these legumes be given every chance to perpetuate themselves. Most pasture legumes in Florida act as annuals, and must rely on a self-reseeding or must be reseeded by the grower. Not only is hand reseeding costly, but it is a practice not readily accepted by growers, so a natural reseeding is preferred.

If more colonies of honeybees in Florida result in the legumes setting a heavier seed crop, and this does appear to be true, then volunteer legume stands will be improved, and as a result pastures and cattle will be benefited.

In view of the preliminary work in 1949, it was decided to

expand the research on legume pollination in 1950. In the early spring of 1950, cages and colonies of bees were set up on Crimson clover fields at Monticello and Quincy, on a White Dutch clover field at Wewahitchka, on Red Clover at Gainesville, and on a Hubam (annual white sweet) clover field near Micanopy.

Cages six feet square and 30 inches high were covered with plastic screen and these were used to exclude bees in one instance and to hold honeybees in the other being placed on uniform stands of the respective clovers. In each case, a staked area was set aside without a cage, and this was accessible to all bees and insects and harvested as a comparison to the caged areas. (Results are given in detail on the following page.)

At Quincy, Crimson Clover produced 3 pounds of seed per acre under cages without bees, 64 pounds with bees under cages, and 105 pounds in the staked area where all bees and insects were present.

The experiment at Monticello on Crimson Clover was conducted in the same manner, but clover seed yields were much lower and it is believed that a shortage of a certain plant nutrient, probably boron, was responsible. Without bees the seed yield was again 3 pounds per acre, 31 pounds with bees, and 10 pounds with all bees and insects. (Records of those experiments on following page show numbers of seedpods per head, seed per head, and percentage of pods with seed.)

The Sweet clover (Hubam) experiment was not considered satisfactory, as most of the clover died under the cages before setting seed. It was thought the cages may have excluded too much light or air movement. Extreme drought also contributed to the failure in this case. It was noted that honeybees were very active on this field of Hubam as well as on several hundred more acres nearby, indicating a preference by the bees for Hubam pollen and nectar.

The Red clover experiment, while not producing as much seed as the previous year, was satisfactory. No Red clover seed was harvested from the cage without bees, while 38 pounds per acre were harvested from the cage with bees, and 40 pounds were harvested from the staked area which had all bees and insects. The experiment on White Dutch clover was destroyed when cattle broke into the clover field and wrecked the cages.

Summarizing the date to date for the two years, the data indicate a definite need for bees in the pollination and seed set of the legumes studied. Probably some areas are already well supplied with bees and insects capable of pollinating legumes, but as the pasture program expands into new areas of the state, there is little doubt but that more honeybees will greatly enhance the chances of success in the cattlemen's grass-legume pasture program.

DATA ON CLOVER POLLINATION FOR 1950

Crimson Clover
North Florida Experiment Station, Quincy

Treatment of 36 sq. ft. plots	Avg. from 20 seed-heads			Yield calculated from 36 sq. ft. of harvested plot
	Pods	Seed	% pods with seed	Pounds of seed per acre
Caged—No Bees	75	0	0	0
Caged—With Bees	77	8	10.4	38
Not Caged—all Bees and insects	76	20	26.3	40

Crimson Clover
Monticello

Treatment of 36 sq. ft. plots	Avg. from 20 seed-heads			Yield calculated from 36 sq. ft. of harvested plot
	Pods	Seed	% pods with seed	Pounds of seed per acre
Caged—No Bees	....	....	....	3
Caged—With Bees	....	....	....	7
Not Caged—all Bees and insects	....	....	....	74

Sweet Clover (Hubam)
Micanopy

Treatment of 36 sq. ft. plots	Avg. from 20 seed-heads			Yield calculated from 36 sq. ft. of harvested plot
	Pods	Seed	% pods with seed	Pounds of seed per acre
Caged—No Bees	69	0	0	3
Caged—With Bees	66	9	13.6	31
Not Caged—all Bees and insects	65	4	6.2	10

Red Clover
Gainesville

Treatment of 36 sq. ft plots	Avg. from 20 seed-heads			Yield calculated from 36 sq. ft. of harvested plot
	Pods	Seed	% pods with seed	Pounds of seed per acre
Caged—No Bees	65	2	3.1	3
Caged—With Bees	72	52	72.2	64
Not Caged—all Bees and insects	71	59	83.1	105

Honey Processing

Honey extracting tanks holding fifty frames each.

The honey combs in their wooden frames are brought into the processing plant. Nine frames fit into the standard box.

First step is taking the frames to the electric knife where the wax is cut off, as well as the surplus comb, so the liquid can be extracted. The bees seal the combs with the wax they produce and each section must be opened to permit the juice to flow.

The frames are put in mechanical extractors which hold 50 at a time. On goes the switch and the whirling motion in the cylinder throws the liquid out of the comb against the steel wall of the cylinder.

The liquid honey flows into another tank and is then pumped through 180 feet of copper coils which are heated to between 140 and 160 degrees.

By this time the honey is liquid and goes through a filter process where bits of wax or impurities are sifted out. Next

Filling tins from 500 gallon stainless steel storage tank.

step in the process is to the storage tank where the liquid gold remains until packaged.

The honey is bottled or put in tins. Large drums of honey are filled for use by commercial bakers. From small jars to 60 lb. tins to drums, all sizes are prepared.

The average production of two extracting tanks is approximately three tons per day.

BEES WAX

The wax when it was cut off the comb went into a container where it was melted over steam filled tubing and flowed into a pan. Beeswax is even more valuable than the honey itself and commands a good price on the market, for it is used for so many things. Some of the wax is sold to processors for use in other commodities and some is sent to a company to convert into new combs.

The manufacturing of a basic comb is one of the most interesting angles of this honey business. Four thin wires are run across a frame. Over this is poured the liquid beeswax to make a comb of about a thirty second of an inch thick. This is then marked with a press which makes the comb the same natural design as though the bees constructed it.

When put in the fields the bees swarm to it and start producing a comb full of honey on the simulated base. Combs are used over and over again. After the honey is extracted the frame, with its comb, is taken out and reused.

Honey is a wholesome natural food.

www.ingramcontent.com/pod-product-compliance
Lightning Source LLC
Chambersburg PA
CBHW032302150426
43195CB00008BA/552